Advanced Flame Retardant Materials

Advanced Flame Retardant Materials

Special Issue Editor

Fouad Laoutid

MDPI • Basel • Beijing • Wuhan • Barcelona • Belgrade • Manchester • Tokyo • Cluj • Tianjin

Special Issue Editor
Fouad Laoutid
Materia Nova Research Center
Belgium

Editorial Office
MDPI
St. Alban-Anlage 66
4052 Basel, Switzerland

This is a reprint of articles from the Special Issue published online in the open access journal *Materials* (ISSN 1996-1944) (available at: https://www.mdpi.com/journal/materials/special_issues/flame_retardant).

For citation purposes, cite each article independently as indicated on the article page online and as indicated below:

LastName, A.A.; LastName, B.B.; LastName, C.C. Article Title. *Journal Name* **Year**, *Article Number*, Page Range.

ISBN 978-3-03928-350-7 (Pbk)
ISBN 978-3-03928-351-4 (PDF)

Cover image courtesy of Fouad Laoutid.

Contents

About the Special Issue Editor

Fouad Laoutid obtained a Masters degree in material science from the University of Montpellier in 1999. He received his Ph.D. degree from the University of Montpellier in 2003 and joined the Materia Nova research center (Belgium) in 2006. After his PhD, he worked as the technical incubator of École des Mines d'Alès in 2003 on the development of fire-resistant polyurethane foam before joining École des Mines d'Alès in 2004 as a lecturer. Dr Laoutid obtained a Habilitation à Diriger des Recherches in polymer chemistry in 2016 at the University of Montpellier. He has over 21 years' experience in polymer formulation of sustainable flame-retardant polymeric materials, bio-based materials, (nano)composites, polymer blends, and recycling.

Preface to "Advanced Flame Retardant Materials"

Fireproofing of combustible polymeric materials is required for several applications to limit the risk of fire development. This task is particularly complex due to the multiple kinds of polymers that have different properties and can be used in many different applications (fabrics, coatings, foams, cables, etc.), having specific functional properties (mechanical, electrical, UV-resistance, etc.) and different fire-retardant requirements. Currently, the most studied fireproofing strategy is aimed at developing solutions to promote the formation of barrier layers during the combustion. The interest in this approach lies in the thought that the formed layer will limit the amount of combustible volatiles and toxic products released during combustion. Phosphorus is currently the most widely used element for developing this type of structure.

In addition to the technical requirements, the assessment of the environmental effects of these flame-retardant materials is becoming increasingly important. The development of flame retardants is therefore two-pronged: technical performance (including toxicity) and environmental impact. This trend has paved the way for the development of a new class of bio-based flame retardants that aim to find new alternatives for the recovery of products from biomass or waste. These additives are mainly used in combination with phosphorous through additive or reactive pathways to increase their char forming ability as well as to enhance the thermal stability of the formed char. Phosphorus can be used as an additive or directly incorporated into the polymeric backbone. Inherent flame-retardant polymers could have many advantages such as limiting the flame retardant agent migration and better transparency.

This document is a collection of 11 articles that were published in the framework of the Special Issue (Advanced Flame Retardant Materials) and will allow the reader to gain a wide view on recent developments in the field of improving the fire behavior of polymeric materials. These articles describe work that covers the aspects mentioned above, ranging from the development of bio-based flame retardants, flame retardancy of biopolymers by additive and reactive routes, development of synergistic effects, understanding the contribution to flammability of different phosphorus groups, and flame retardancy of thermoset materials and fabrics.

Fouad Laoutid
Special Issue Editor

Article

Preparing the Degradable, Flame-Retardant and Low Dielectric Constant Nanocomposites for Flexible and Miniaturized Electronics with Poly(lactic acid), Nano ZIF-8@GO and Resorcinol Di(phenyl phosphate)

Mi Zhang, Yu Gao, Yixing Zhan, Xiaoqing Ding, Ming Wang and Xinlong Wang *

School of Chemical-Engineering, Nanjing University of Science & Technology, Nanjing 210094, China; 18120166132@163.com (M.Z.); gaoyu@njust.edu.cn (Y.G.); 18751851021@163.com (Y.Z.); ding1771556282@163.com (X.D.); m18751966157@163.com (M.W.)
* Correspondence: wxinlong323@163.com; Tel.: +86-25-8431-5949

Received: 12 August 2018; Accepted: 4 September 2018; Published: 18 September 2018

Abstract: Degradable, flame retardant, and flexible nanocomposite films with low dielectric constant were prepared with poly (lactic acid) (PLA), nano ZIF-8@GO, and degradable flame-retardant resorcinol di(phenyl phosphate) (RDP). The SEM results of the fractured surfaces indicated that ZIF-8@GO and RDP were dispersed uniformly in the PLA matrix. The prepared films had good mechanical properties and the tensile strength of the film with 1.5 wt% of ZIF-8@GO was increased to 48.2 MPa, compared with 38.5 MPa of pure PLA. Meanwhile, the nanocomposite films were flexible due to the toughing effect of RDP. Moreover, above 27.0% of limited oxygen index (LOI) and a VTM-0 rating were achieved for the nanocomposite films. The effects of nano ZIF-8@GO hybrids and RDP on the dielectric properties were investigated, and the results showed that ZIF-8@GO and RDP were beneficial in reducing the dielectric constant and dielectric loss of the nanocomposites.

Keywords: ZIF-8@GO hybrids; PLA; dielectric constant; flame-retardant; flexible

---

1. Introduction

In recent years, electronic products such as mobile phones have been updated faster, which results in a large volume of e-waste and poses a risk to the sustainable environment [1]. It is estimated that globally, about 20–50 million tons of waste electronic equipment is discarded annually [2]. Green electronic products made of environmentally friendly and disposable materials, such as degradable polymers, are ideal candidates to resolve the pollution from electronic waste [3,4].

Owing to the miniaturization, portability, and flexibility requirements of electronic products, polymeric materials with low dielectric constant and flexibility are getting more attention in integrated circuit devices as interlayer dielectrics [5–7]. By reducing the dielectric constant of the dielectric material used in the integrated circuit, the leakage current of the integrated circuit, the capacitance effect between the wires, as well as the heating of the integrated circuit can be reduced efficiently [8]. Another important problem when thinking of the application of polymer-based low dielectric materials, is that most polymeric materials are easy to burn [9]. To use electronic equipment safely, it is essential that the polymer-based dielectric layer should have flame retardant properties.

Nanocomposites exhibit improved physical properties, compared to those of the unfilled polymer matrix, i.e., Mechanical strength [10,11], dielectric properties [12,13], thermal properties [14], and fire resistance [15,16], etc. Oliviero et al. [17] studied the dielectric properties of sustainable nanocomposites, based on zein protein and lignin for biodegradable insulators. Santanu Singha et al. [18] investigated the effects of nano ZnO on the dielectric properties of epoxy nanocomposites and found that the volume fraction and nature of the interfaces in most of the composites, influence the dielectric properties of the nanocomposites. Rao [19] investigated the structural and electrical properties of novel polyvinyl alcohol (PVA)-CuO nanocomposite films, and the results showed that the dielectric constant was reduced with an increase in both frequency and CuO concentration; and the dielectric loss was increased with increase in frequency and decreased with increase in CuO concentration. Regarding the flame retarded properties of nanocomposites, Hapuarachchi [20] developed the poly(lactic acid) (PLA) nanocomposites with improved flame retardancy, utilizing the unique properties of sepiolite nanoclay and multi-7walled nanotubes. Li groups [21] developed the core-shell nanostructured MWCNT-DOPO-OH through a three-step process and added them to aluminum hypophosphite/poly(lactic acid) (AHP/PLA) flame retardant systems, to improve both flame retardancy and mechanical properties. The results indicated that the PLA nanocomposites with 1 wt% MWCNT-DOPO-OH and 14 wt% AHP, achieved a UL 94 V-0 rating and limiting oxygen index (LOI) value of 28.6%. Ye et al. [22] reported that the addition of organically modified montmorillonite (OMMT) and aluminium diethylphosphinate (AlPi) into the PLA matrix could promote char-forming and suppress the melt dripping. However, few papers have reported on the preparation of degradable nanocomposites with good flame retardancy and low dielectric constant, simultaneously.

In this paper, PLA, nano-ZIF-8@GO hybrids of nano zeolite imidazole frameworks (nano-ZIF-8) and graphene oxide (GO), and resorcinol di(phenyl phosphate) (RDP) were used to prepare the nanocomposites. PLA, a kind of biodegradable polyester produced from renewable resources, is now increasingly viewed as a valuable biosourced polymer alternative in applications, such as electronics [23,24]. The hybrids combine the unique advantage of metal–organic frameworks (MOFs) and GO, such as crystalline and highly ordered structures, ultrahigh porosity and large surface area, two-dimensional structure with rich carboxyl, hydroxyl, and epoxy groups [25]. The flame-retardant RDP is a biodegradable polymeric compound with good thermal stability, as shown in Scheme 1. The effects of nano ZIF-8@GO hybrids and RDP on dielectric behavior, mechanical properties, and flame retardancy of the PLA nanocomposite films have been systematically studied.

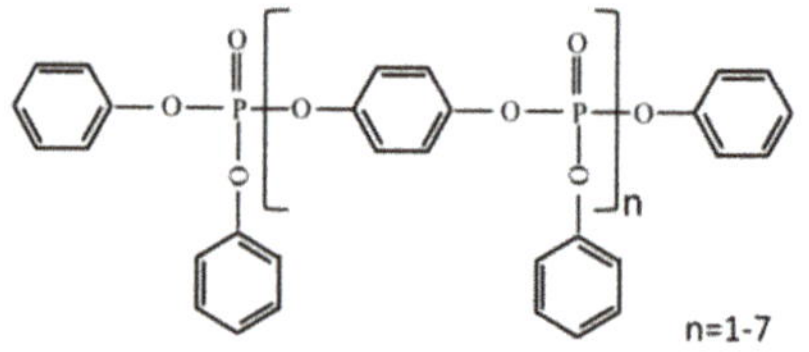

Scheme 1. The chemical formula of resorcinol di(phenyl phosphate) (RDP).

2. Experimental

2.1. Materials

Poly (lactic acid) (PLA 290, M_W = 50,000−60,000, D = 1.4–1.5, content of D-lactide = 0.5%) was obtained from Haizheng Biological Materials Co., Ltd., Taizhou, China. Resorcinol bi(diphenyl phosphate) (RDP) was provided by Wansheng New Material Co., Ltd., Linhai, China. 2-Methyl imidazole (98%) was purchased from Aladdin Industrial Corporation, Shanghai (China). $Zn(NO_3)_2 \cdot 6H_2O$ was supplied by Xilong Chemical Co., Ltd., Shantou, China. Deionized water was produced in our lab. Methanol (CH_3OH), graphite powders, phosphorus pentoxide (P_2O_5), potassium persulfate ($K_2S_2O_5$), and hydrogen peroxide (H_2O_2) were provided by Sinopharm Chemical Reagent Co., Ltd., Shanghai, China. Concentrated sulfuric acid (H_2SO_4, 95~98%), Chloroform ($CHCl_3$, 99%), potassium permanganate ($KMnO_4$), and hydrochloric acid (HCl) were obtained from Lingfeng Chemical Reagent Co., Ltd., Shanghai, China.

2.2. Preparation of ZIF-8@GO

Graphene oxide (GO) was prepared with an improved Hummers method [26]; 1.602 g of 2-methyl imidazole and 1.487 g of $Zn(NO_3)_2 \cdot 6H_2O$ were dissolved in 100 mL of methanol (CH_3OH), respectively, and 0.163 g of GO was dispersed in 100 mL of CH_3OH. Then, the GO suspension was added to the above solution with stirring. Finally, the reaction mixture was stirred for 1h at room temperature and ZIF-8@GO particles were centrifuged, washed with CH_3OH, and then dried at 80 °C for 48 h.

2.3. Preparation of the Nanocomposite Films

First, the dried PLA was dissolved in 40 mL of chloroform ($CHCl_3$) with magnetic stirring for 2 h. An appropriate amount of ZIF-8@GO and RDP was dispersed in 20 mL of $CHCl_3$ by sonication to form a uniform dispersion solution, and then poured into the PLA solution. The mixture was stirred for 4 h and kept for 1 h. The mixed solution was then casted into a film using an automatic coater (MRXTMH250, Mingruixiang Automation Equipment Co., Ltd., Shenzhen, China). After the solvent was evaporated at room temperature, the film was dried in an oven at 50 °C for 72 h, to further remove the residual solvent. The concrete formulations of nanocomposite films are listed in Table 1.

Table 1. Formulations and combustion tests of nanocomposite films.

Samples	PLA (wt%)	RDP (wt%)	ZIF-8@GO (wt%)	LOI (%)	Rating
PLA	100	0	0	21.0 ± 0.2	NR [b]
PLA/RDP	91.0	9.0	0	30.0 ± 0.3	VTM-0
PLA-1	90.4	9.0	0.6	29.3 ± 0.5	VTM-0
PLA-2	90.1	9.0	0.9	28.5 ± 0.2	VTM-0
PLA-3	89.8	9.0	1.2	27.8 ± 0.6	VTM-0
PLA-4	89.5	9.0	1.5	27.0 ± 0.4	VTM-2

NR [b]: no rating.

2.4. Measurement and Characterization

Transmission electron microscope (TEM, JEM-2100, Tokyo, Japan) was operated to observe the morphologies of ZIF-8@GO.

X-ray diffraction (XRD, Bruker D8 Advance diffractometer, Karlsruhe, Germany) measurements were performed at 40 kV and 40 mA, with Cu K_α radiation (0.15418 nm).

Fourier transform infrared spectroscopy (FTIR) spectra were recorded on a FTIR-8400S spectrometer (Shimadzu, Japan), at a range of 400–4000 cm^{-1}, with a resolution of 4 cm^{-1}, to observe the structure of PLA nanocomposites.

Scanning electron microscope (SEM, Hitachi S-4800, Tokyo, Japan) with an Energy Dispersive Spectrometer (EDS, GENESIS 2000, EDAX, Mahwah, NJ, USA) was employed to observe the fracture morphologies of PLA nanocomposite film. All samples were coated with gold before the test.

Tensile testing measurements were performed on a CMT tensile tester (Sans, Shenzhen, China), at a rate of 10 mm/min at room temperature, referred to ISO 1184-1983 (GB 13022-1991, China). The sample values were averaged by five measurements and sample size was (150 mm $\pm$ 2 mm) $\times$ (20 mm $\pm$ 2 mm). The thickness of the sample was 0.15 mm.

Limiting oxygen index (LOI) tests were measured according to ASTM Standard D2863-97, with a JF-3 oxygen index meter from Jiangning Analysis Instrument Co., Ltd. (Nanjing, China). The tests were performed five times, and sample size was (200 mm $\pm$ 5 mm) $\times$ (50 mm $\pm$ 2 mm).

Vertical burning tests were performed according to ISO 9773 using a vertical burning tester (CZF-3, Jiangning Analytical Instrument Co., Ltd., Nanjing, China), with a standard of UL-94. The sample values were measured five times, and sample size was (200 mm $\pm$ 5 mm) $\times$ (50 mm $\pm$ 2 mm). The thickness of the sample was 0.15 mm.

Thermogravimetric analysis (TGA) was carried out from 35 °C to 600 °C using a DTG-60 thermoanalyzer instrument (Shimadzu, Tokyo, Japan), at 20 °C/min under nitrogen. The mass of the samples was kept within 3–5 mg in an alumina crucible.

The dielectrical characterization of the nanocomposites was performed using a high-frequency LCR meter (TH2816, L = 100 µH, K = 6, DZ5001, Nanjing, China), with a frequency range from 1 kHz to 150 kHz, as described in Reference [27]. In the experimental process, air was used as a reference. The permittivity (ε_r) was calculated from the formula as follows:

$$\varepsilon_r = 14.4 \cdot (C_1 - C_2) \cdot d / \Phi^2 \tag{1}$$

where C_1, C_2 is the capacitor of air and sample (pF), and d is the thickness of the film (cm). Φ is the diameter of the film (cm) and the general value of Φ was 3 cm. The dielectric loss tangent ($tan\delta$) was calculated from the formula as follows:

$$tan\delta = C_1 \cdot (Q_1 - Q_2)/(C_1 - C_2) \cdot Q_1 \cdot Q_2 \tag{2}$$

where Q_1, Q_2 is the quality factor of air and sample, respectively.

3. Results and Discussion

3.1. Characterization of ZIF-8@GO

The morphologies of ZIF-8@GO are shown in Figure 1a. It is apparent that many ZIF-8 nanoparticles were anchored on the surface of GO sheets by chelation between oxygen-containing functional groups on GO and Zn ions in ZIF-8 [28]. Figure 1b shows the XRD patterns of ZIF-8, GO, and ZIF-8@GO. The peak centered at 10.9° for GO, corresponds to the reflections of the (002) plane [29]. The XRD pattern of the synthesized ZIF-8 was similar to that reported in previous literature [30]. The XRD pattern of ZIF-8@GO showed the diffraction peaks of ZIF-8. However, the (002) reflection of GO was absent because the incorporation of ZIF-8 had destroyed the regular stack of GO. This phenomenon was inconsistent to those reported by others in References [31,32]. The XRD analysis further demonstrated that ZIF-8 nanoparticles had been effectively loaded onto the GO sheets.

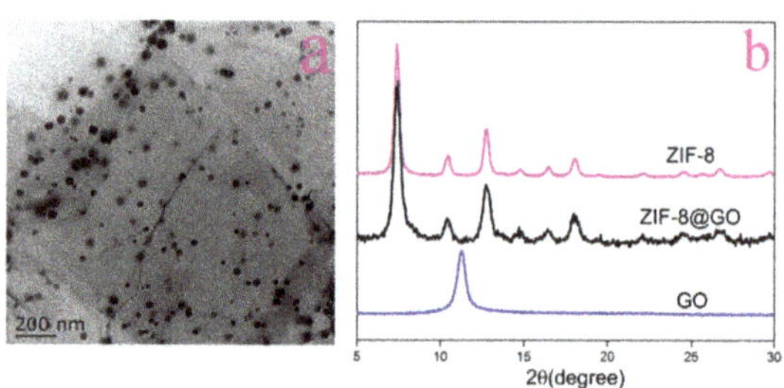

Figure 1. (a) Transmission electron microscope (TEM); (b) X-ray diffraction (XRD) of ZIF-8@GO.

3.2. SEM and Transparency of the Nanocomposite Films

Figure 2a shows the SEM image of pure PLA. As can be seen, pure PLA exhibited a fractured morphology with some cracks, which is the characteristic of rigid and fragile materials. The fractured morphology of PLA-2 with 0.9 wt% ZIF-8@GO and 9.0 wt% RDP is shown in Figure 2b,c, and the surface with no large agglomerations is observed, indicating that ZIF-8@GO and RDP are dispersed uniformly in the PLA matrix. RDP is a kind of organic polyphosphate and is miscible with PLA [33]. Regarding the good dispersion of ZIF-8@GO in PLA, the strong interactions between the organic linkers in ZIF-8 and PLA chains provide good affinity for ZIF-8@GO with PLA. Moreover, ZIF-8 nanoparticles anchored on GO sheets prevented GO from stacking together and made them disperse well in the PLA matrix. Moreover, the surface of ZIF-8@GO may be modified by the organic RDP, since many P=O and P–O–C groups in RDP can form hydrogen bonds with the –OH groups existing on the surface of GO sheets, or coordination bonds with Zn ions in ZIF-8. Thus, RDP molecules tend to be absorbed on the surface of ZIF-8@GO nanoparticles, which enhances the compatibility between ZIF-8@GO nanoparticles and the PLA matrix. The EDS of PLA-2 nanocomposite shown in Figure 2d, further proves the presence of ZIF-8@GO and RDP in the nanocomposites. In Figure 2e, all the films are found to be transparent enough to see "ABC" letters on paper beneath the films clearly, even though the content of ZIF-8@GO reaches 1.5 wt%.

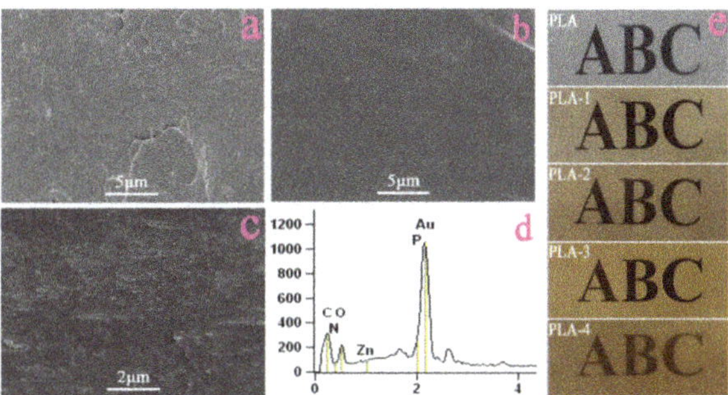

Figure 2. (**a**) Scanning electron microscope (SEM) of pure poly (lactic acid) (PLA); (**b**,**c**) SEM of PLA-2; (**d**) EDS of PLA-2; (**e**) Optical images of the prepared films.

3.3. Mechanical Properties of the Nanocomposite Films

The typical stress-strain curves for the nanocomposite films are shown in Figure 3a. The linear stress-strain behavior of films in the initial region was observed as the 'Hookean' region, which was due to the changes in bond angles and spacing of PLA [34]. The rest of the region was considered as the plastic region, which is attributed to the movement of the PLA chain under the external force. The related data of tensile strength from the stress-strain curve, is illustrated in Figure 3b. It is well known that PLA is a rigid thermoplastic polymer with 38.5 MPa of tensile strength; however, the tensile strength of PLA/RDP film is only 34.7 MPa, which may be ascribed to the fact that the RDP molecules inserts PLA chains and increases the spacing of PLA chains, weakening the inter-chain stress during the blending process, resulting in reduced tensile strength [35]. The nanocomposite films show a trend of increase in tensile strength, for example, when the content of ZIF-8@GO was 1.2 wt%, the tensile strength of PLA-3 was 47.4 MPa, which was an increase of 22.8% compared with the pure PLA. This may be owing to the fact that the added ZIF-8@GO particles played a reinforcing effect for PLA, where the Zn-containing groups of ZIF-8@GO could interact with the ester groups of PLA to form a partial crosslinking through coordination interaction, as exhibited in Figure 4a. In Figure 4a, the Zn-containing groups of ZIF-8@GO have open sites, as described in Reference [36], which have a strong coordination effect with C=O and C–O–C of ester groups in PLA. To support this conclusion, the FTIR curves of PLA and PLA-4 as a representative are exhibited in Figure 4b. The peak at 1761 cm^{-1} was attributed to the stretching vibration of C=O in PLA [8]. While adding ZIF-8@GO, the peak shifted to 1756 cm^{-1} for PLA-4. Similarly, the peak at 1091 cm^{-1}, corresponding to the C–O–C vibration in PLA shifts the lower wavenumber. The shifts of these peaks could be due to the strong interaction between PLA and ZIF-8@GO. Moreover, RDP, as a compatibility agent, promotes the uniform distribution of ZIF-8@GO in PLA, which is beneficial to the improvement of mechanical properties [37,38]. The digital photo for the PLA films in Figure 3b, shows that the prepared nanocomposite films are so flexible, which could be due to plasticity of RDP for PLA, as well as the toughing effects of ZIF-8@GO in PLA [39]. The elongations at break of PLA-1, PLA-2, and PLA-3 are about 11.48%, 15.17%, and 16.33%, respectively, compared with 9.92% for PLA.

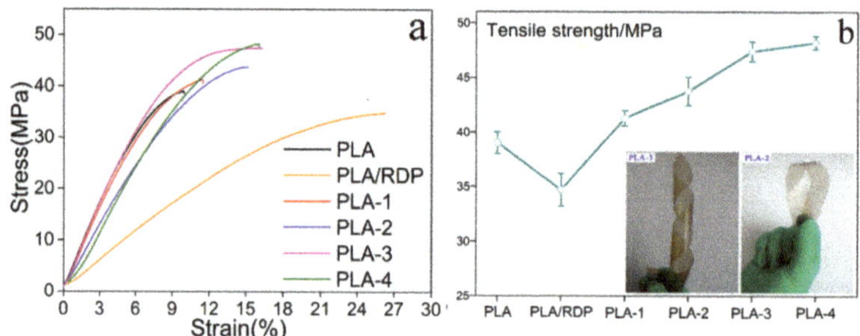

Figure 3. The stress-strain curves (**a**) and tensile strength (**b**) of the nanocomposite films.

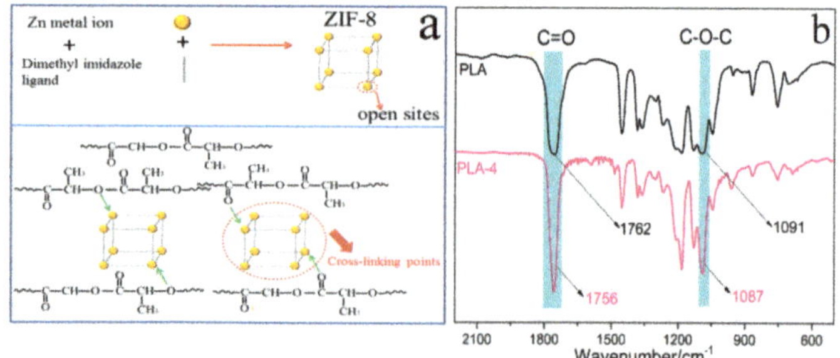

Figure 4. The enhancement mechanism (**a**) and Fourier transform infrared spectroscopy (FTIR) curves (**b**) the nanocomposite films.

3.4. Flame Retardant Properties of the Nanocomposite Films

Table 1 shows the results of the combustion performance for the nanocomposite films, including the limit oxygen index (LOI) and vertical burning rating. The LOI is defined as the minimum oxygen concentration required for combustion under the oxygen-nitrogen atmosphere. It is generally believed that pure PLA has a poor flame-retardance property, as its LOI is only 21.0%. In addition, PLA burns vigorously with serious droplet and no residues, which will lead to the spread of flames and secondary combustion hazards, as reported previously in Reference [40]. The incorporation of RDP can remarkably improve the flame retardancy of PLA, with a LOI of 31.0% and VTM-0 rating. With respect to the flame-retardant mechanism, the organophosphorus flame retardants often play an effective role on flame retardancy, both in condensed and gas phase. In condensed phase, they decompose into phosphoric acid and pyrophosphate, which can catalyze charring process to form the char layer on polymer. In gas phase, they can react with the degrading polymer to quench radicals or produce PO· radicals, which can react with the H· or OH·, and thus act as a fire retardant. As shown in Figure 5a, the amounts of char residues of the nanocomposites are poor, suggesting RDP cannot promote the carbonization of PLA. Therefore, we speculate that the gas phase mechanism is mainly responsible for the improvement of flame retardancy, for the prepared nanocomposites. Fang has proved the statement that some organophosphorus flame retardants are mainly active in the gas phase and not in the condensed phase using TG-FTIR [41]. The addition of ZIF-8@GO into the PLA films results in a slightly lower LOI, but remains in high values. For instance, when the ZIF-8@GO content is 1.5 wt%, the LOI of PLA-4 is 27.0%. The reduction in flame retardance may be attributed to ZIF-8@GO nanoparticles, which catalyzes the decomposition of RDP at low temperature regions, as shown in Figure 5. In Figure 5a, the initial decomposition temperatures of PLA and PLA/RDP are about 345.5 °C and 340.9 °C. While adding the RDP and ZIF-8@GO, the initial decomposition temperatures of PLA-2 and PLA-4 nanocomposites were about 295.1 °C and 264.4 °C, which were lower than 340.9 °C for PLA/RDP. Moreover, the TG curves of RDP and the RDP/ZIF-8@GO blend are obtained in Figure 5b. The initial decomposition temperatures of RDP and RDP/ZIF-8@GO blend were 300.4 °C and 265.0 °C, respectively. The RDP/ZIF-8@GO blend decomposed earlier than that of RDP, which proved the catalytic effect of ZIF-8@GO further [42,43]. The decomposition of RDP at low temperature impairs the effect of flame retardancy of RDP. The SEM of char residue after combustion of the nanocomposites (PLA-2 film), is shown in Figure 6. Although this char layer is continuous and beneficial to improvement of flame retardance of the PLA nanocomposites, the amount of char residue after combustion of the nanocomposites is so poor that it cannot improve flame retardancy effectively in condensed phase.

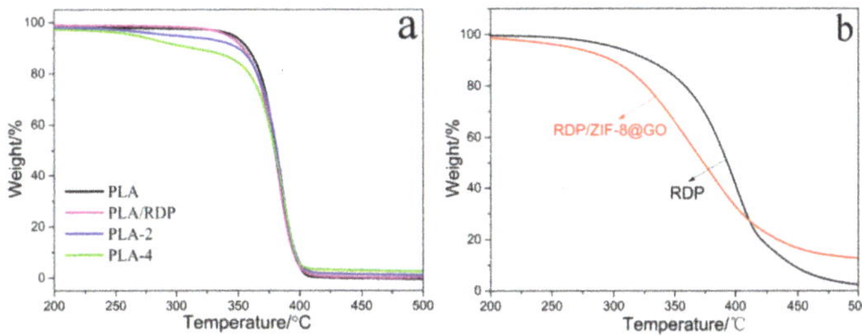

Figure 5. TG curves of PLA nanocomposite films (**a**); RDP and RDP/ZIF-8@GO blend (**b**).

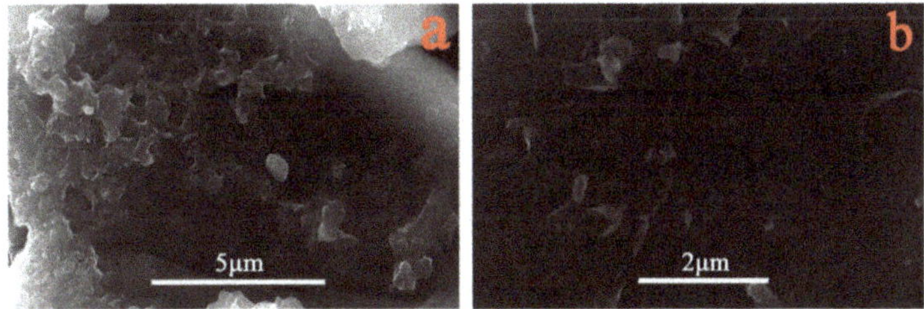

Figure 6. SEM images of PLA-2 nanocomposite films after combustion under different resolutions. (**a**) 5 μm; (**b**) 2 μm.

3.5. Dielectric Properties of the Nanocomposite Films

The dielectric constant and dielectric loss tangent are the most important parameters for dielectric materials. The dielectric constant and dielectric loss tangent of the nanocomposite films from 1 kHz to 150 kHz have been tested, and the results are shown in Figure 7. As shown in Figure 7a, the dielectric constant show low-frequency dependence within the measuring frequency. When the frequency is less than 30 kHz, the dielectric constant decreases rapidly with the frequency. This is because the dielectric constant of the PLA films, depends on their abilities to polarize at a given frequency. There are four possible polarizations that could contribute to the dielectric behavior: Electronic, ionic, dipolar, and interfacial polarization [44]. At low frequencies, all four types of polarization have enough time to occur, which is helpful to reduce the dielectric constant. As the frequency increases, the contributions of interfacial, dipolar, and ionic polarizations do not have enough time to adjust to the change of frequency and become ineffective, only the electronic polarization still plays a role in the dielectric constant [8,45]. The decrease of dielectric permittivity with increasing frequency is due to the relaxation behaviors [46]. Figure 7b shows the dielectric loss tangent for the films from 1 kHz to 150 kHz. In the range of the discussed frequency, the change of dielectric loss is not obvious. It is seen that the values of all the films are in a range from 0.045 to 0.070.

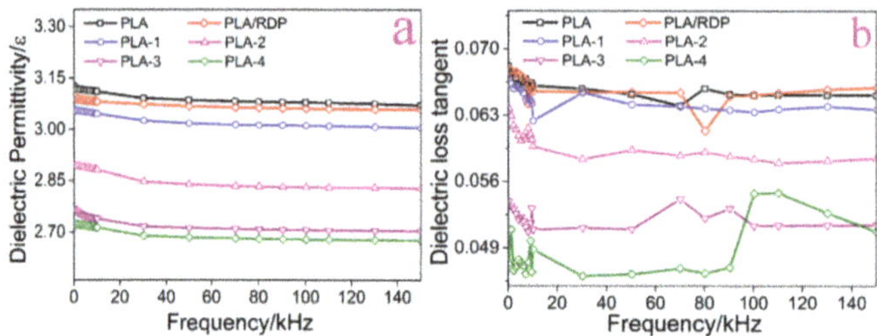

Figure 7. Dielectric permittivity (**a**) and dielectric loss tangent (**b**) of the films.

The effects of ZIF-8@GO and RDP on the dielectric properties of the nanocomposites are shown in Figure 8. While only adding RDP, the dielectric constant decreases slightly compared with the pure PLA, which may be ascribed to the presence of several benzene rings in RDP, because it is well-known that the benzene ring has a very low dipole moment; for example, the dielectric constant of polystyrene is about 2.4 at 10^3 kHz [47,48]. With addition of ZIF-8@GO, the dielectric constant had a rapid decline and achieved the lowest value at 1.5 wt% of ZIF-8@GO. For example, the dielectric constant of PLA-4 reached 2.61, compared with 3.1 for pure PLA at 10 kHz. The decrease of the dielectric constant is attributed to the good dispersion of fillers [49], nanoparticle effect of ZIF-8@GO, as well as the interface effect [46], as shown in Figure 9. ZIF-8 with a nano porous structure increased the porosity density in the PLA matrix, thereby reducing the polarization molecular density [50]. GO, with the special two-dimensional planar structure, lessened the efficiency of molecular piling and increased the free volume of the PLA, thereby diluting the density of polarized molecules in the PLA matrix [51]. The presence of GO also created a significant enhancement in porosity, owing to the formation of new pores at the interface of GO and ZIF-8 particles, as exhibited in Figure 9 [52]. Furthermore, as shown in Figure 9, the PLA molecules and the ZIF-8@GO nanoparticles in both the first layer and the second layer of the interface multi-core region have a strong interaction, which restricts end-chain or side-chain movement. The interaction has a profound effect on the dielectric behavior at low frequency [53,54]. However, the collaborative effect between ZIF-8@GO and RDP on the dielectric constant needs to be further investigated. Figure 8b shows the change of the dielectric loss tangent, as a function of the addition of ZIF-8@GO. It was observed that the dielectric loss decreases with an increase in the ZIF-8@GO content. The phenomenon could be explained as the Coulomb blockade effect of ZIF-8@GO. The ZIF-8@GO nanoparticles could cause a high charging energy for the tunneling electrons and inhibit the charge transfer through the whole system from migrating directionally, reducing the conduction loss which represents the flow of charge through the dielectric materials [55,56].

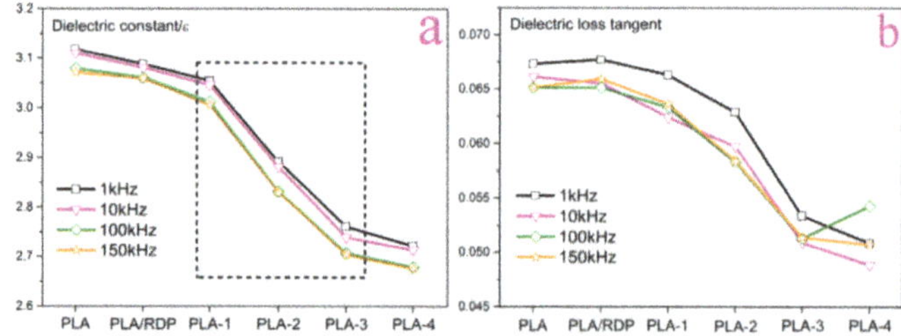

Figure 8. Dielectric permittivity (**a**) and dielectric loss tangent (**b**) with the content of ZIF-8@GO.

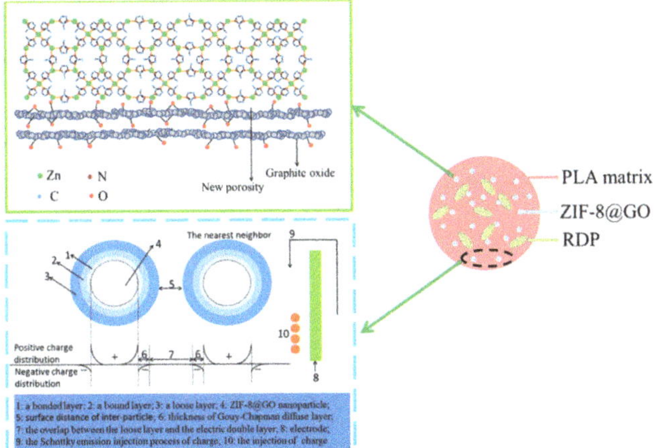

Figure 9. Mechanism of the dielectric permittivity reduction for the films. (**a**) Nanoparticle effect; (**b**) Interface effect [57].

4. Conclusions

In conclusion, ZIF-8@GO hybrids were synthesized and characterized by TEM and XRD. The ZIF-8@GO particles were homogenously dispersed in the PLA matrix due to the action of RDP, as well as the interfacial interaction between PLA and ZIF-8@GO. The tensile strength was improved, compared with pure PLA, due to the reinforcing effect of ZIF-8@GO. The nanocomposite films showed good flame retardance. The LOI of the PLA-2 film was 28.5% and a VTM-0 rating was obtained. The dielectric constant of PLA films was decreased, owing to the nanoparticle and interface effects. While adding 9.0 wt% RDP and 1.5 wt% ZIF-8@GO, the dielectric constant of PLA-4 reached 2.61, compared to about 3.30 and 4.40 for PI and EP at 1 kHz frequency, respectively. Moreover, the dielectric loss tangent had a general trend of slightly decreasing with the addition of ZIF-8@GO, which could be explained as the Coulomb blockade effect of ZIF-8@GO.

Author Contributions: Conceptualization, X.W.; Methodology, M.Z.; Validation, M.Z.; Formal Analysis, Y.Z. and M.W.; Investigation, M.Z., X.W. and Y.G.; Resources, Y.Z. and X.D.; Data Curation, X.D.; Writing-Original Draft Preparation, M.Z. and X.W.; Writing-Review & Editing, X.W.

Funding: This research was funded by Science and Technology Support Program (Social Development) of Jiangsu Province of China (BE 2013714) and a Project Funded by the Priority Academic Program Development of Jiangsu Higher Education Institutions (PAPD).

Conflicts of Interest: The authors declare no conflict of interest.

References

1. Tansel, B. From electronic consumer products to e-wastes: Global outlook, waste quantities, recycling challenges. *Environ. Int.* **2016**, *98*, 35–45. [CrossRef] [PubMed]
2. Ongondo, F.O.; Williams, I.D.; Cherrett, T.J. How are WEEE doing? A global review of the management of electrical and electronic wastes. *Waste. Manag.* **2011**, *31*, 714–730. [CrossRef] [PubMed]
3. Hwang, S.W.; Song, J.K.; Huang, X.; Cheng, H.; Kang, S.K.; Kim, B.H.; Kim, J.H.; Yu, S.; Huang, Y.; Rogers, J.A. High-performance biodegradable/transient electronics on biodegradable polymers. *Adv. Mater.* **2014**, *26*, 3905–3911. [CrossRef] [PubMed]
4. Irimia-Vladu, M.; Głowacki, E.D.; Voss, G.; Bauer, S.; Sariciftci, N.S. Green and biodegradable electronics. *Mater. Today* **2012**, *15*, 340–346. [CrossRef]
5. Makris, I.; Manteuffel, D.; Seager, R.D. Miniaturized reconfigurable UWB antennas for the integration into consumer electronic products. In Proceedings of the 2nd European Conference on Antennas and Propagation, Edinburgh, UK, 1–16 November 2007; pp. 1–16.

6. Qian, K.; Tay, R.Y.; Lin, M.F.; Chen, J.; Li, H.; Lin, J.; Wang, J.; Cai, G.; Nguyen, V.C.; Teo, E.H.; et al. Direct observation of indium conductive filaments in transparent, flexible, and transferable resistive switching memory. *ACS Nano* **2017**, *11*, 1712–1718. [CrossRef] [PubMed]

7. Cao, Q.; Kim, H.S.; Pimparkar, N.; Kulkarni, J.P.; Wang, C.; Shim, M.; Roy, K.; Alam, M.A.; Rogers, J.A. Medium-scale carbon nanotube thin-film integrated circuits on flexible plastic substrates. *Nature* **2008**, *454*, 495–500. [CrossRef] [PubMed]

8. Gross, T.S.; Soucy, K.G.; Andideh, E.; Chamberlin, K. Detection of plasma-induced, nanoscale dielectric constant variations in carbon-doped CVD oxides by electrostatic force microscopy. *J. Phys. D Appl. Phys.* **2002**, *35*, 723–728. [CrossRef]

9. Tai, C.M.; Li, R.K.Y. Studies on the impact fracture behaviour of flame retardant polymeric material. *Mater. Des.* **2001**, *22*, 15–19. [CrossRef]

10. Wang, X.; Wang, L.; Su, Q.; Zheng, J. Use of unmodified SiO_2, as nanofiller to improve mechanical properties of polymer-based nanocomposites. *Compos. Sci. Technol.* **2013**, *89*, 52–60. [CrossRef]

11. Kashi, S.; Gupta, R.K.; Kao, N.; Hadigheh, S.A.; Bhattacharya, S.N. Influence of graphene nanoplatelet incorporation and dispersion state on thermal, mechanical and electrical properties of biodegradable matrices. *J. Mater. Sci. Technol.* **2018**, *6*, 1026–1034. [CrossRef]

12. Feller, J.F.; Bruzaud, S.; Grohens, Y. Influence of clay nanofiller on electrical and rheological properties of conductive polymer composite. *Mater. Lett.* **2004**, *58*, 739–745. [CrossRef]

13. Kashi, S.; Gupta, R.K.; Baum, T.; Kao, N.; Bhattacharya, S.N. Morphology, electromagnetic properties and electromagnetic interference shielding performance of poly lactide/graphene nanoplatelet nanocomposites. *Mater. Des.* **2016**, *95*, 119–126. [CrossRef]

14. Gu, L.; Qiu, J.; Sakai, E. Effect of DOPO-containing flame retardants on poly(lactic acid): Non-flammability, mechanical properties and thermal behaviors. *Chem. Res. Chin. Univ.* **2017**, *33*, 143–149. [CrossRef]

15. Zhou, K.; Gao, R. The influence of a novel two dimensional graphene-like nanomaterial on thermal stability and flammability of polystyrene. *J. Colloid Interface Sci.* **2017**, *500*, 164–171. [CrossRef] [PubMed]

16. González, A.; Dasari, A.; Herrero, B.; Plancher, E.; Santarén, J.; Esteban, A.; Lim, S.H. Fire retardancy behavior of PLA based nanocomposites. *Polym. Degrad. Stab.* **2012**, *97*, 248–256. [CrossRef]

17. Oliviero, M.; Rizvi, R.; Verdolotti, L.; Iannace, S.; Naguib, H.E.; Maio, E.D.; Neitzert, H.C.; Landi, G. Dielectric properties of sustainable nanocomposites based on zein protein and lignin for biodegradable insulators. *Adv. Funct. Mater.* **2017**, *27*, 1605142. [CrossRef]

18. Singha, S.; Thomas, M.J. Influence of filler loading on dielectric properties of epoxy-ZnO nanocomposites. *IEEE Trans. Dielectr. Electr. Insul.* **2009**, *16*, 531–542. [CrossRef]

19. Rao, J.K.; Raizada, A.; Ganguly, D.; Mankad, M.M.; Satayanarayana, S.V.; Madhu, G.M. Investigation of structural and electrical properties of novel CuO–PVA nanocomposite films. *J. Mater. Sci.* **2015**, *50*, 7064–7074. [CrossRef]

20. Hapuarachchi, T.D.; Peijs, T. Multiwalled carbon nanotubes and sepiolite nanoclays as flame retardants for polylactide and its natural fibre reinforced composites. *Compos. Part A Appl. Sci. Manuf.* **2010**, *41*, 954–963. [CrossRef]

21. Gu, L.; Qiu, J.; Yao, Y.; Sakai, E.; Yang, L. Functionalized MWCNTs modified flame retardant PLA nanocomposites and cold rolling process for improving mechanical properties. *Compos. Sci. Technol.* **2018**, *161*, 39–49. [CrossRef]

22. Ye, L.; Ren, J.; Cai, S.Y.; Wang, Z.G.; Lia, J.B. Poly(lactic acid) nanocomposites with improved flame retardancy and impact strength by combining of phosphinates and organoclay. *Chin. J. Polym. Sci.* **2016**, *34*, 785–796. [CrossRef]

23. Brzeziński, M.; Biela, T. Polylactide nanocomposites with functionalized carbon nanotubes and their stereocomplexes: A focused review. *Mater. Lett.* **2014**, *121*, 244–250. [CrossRef]

24. Raquez, J.M.; Habibi, Y.; Murariu, M.; Dubois, P. Polylactide (PLA)-based nanocomposites. *Prog. Polym. Sci.* **2013**, *38*, 1504–1542. [CrossRef]

25. Fu, Y.; Liu, L.; Zhang, J. Manipulating dispersion and distribution of graphene in PLA through novel interface engineering for improved conductive properties. *ACS Appl. Mater. Interfaces* **2014**, *6*, 14069–14075. [CrossRef] [PubMed]

26. Poh, H.L.; Šaněk, F.; Ambrosi, A.; Zhao, G.; Sofer, Z.; Pumera, M. Graphenes prepared by Staudenmaier, Hofmann and Hummers methods with consequent thermal exfoliation exhibit very different electrochemical properties. *Nanoscale* **2012**, *4*, 3515–3522. [CrossRef] [PubMed]

27. Koteswararao, J. Influence of cadmium sulfide nanoparticles on structural and electrical properties of polyvinyl alcohol films. *Express Polym. Lett.* **2016**, *10*, 883–894. [CrossRef]

28. Kumar, R.; Jayaramulu, K.; Maji, T.K.; Rao, C.N. Hybrid nanocomposites of ZIF-8 with graphene oxide exhibiting tunable morphology, significant CO_2 uptake and other novel properties. *Chem. Commun.* **2013**, *49*, 4947–4949. [CrossRef] [PubMed]

29. Yuan, B.; Wang, B.; Hu, Y.; Mu, X.; Hong, N.; Liew, K.M.; Hu, Y. Electrical conductive and graphitizable polymer nanofibers grafted on graphene nanosheets: Improving electrical conductivity and flame retardancy of polypropylene. *Compos. Part A Appl. Sci. Manuf.* **2016**, *84*, 76–86. [CrossRef]

30. Jiang, H.L.; Liu, B.; Akita, T.; Haruta, M.; Sakurai, H.; Xu, Q. Au@ZIF-8: CO oxidation over gold nanoparticles deposited to metal-organic framework. *J. Am. Chem. Soc.* **2009**, *131*, 11302–11303. [CrossRef] [PubMed]

31. Xiang, Q.; Yu, J.; Jaroniec, M. Enhanced photocatalytic H_2-production activity of graphene-modified titania nanosheets. *Nanoscale* **2011**, *3*, 3670–3678. [CrossRef] [PubMed]

32. Fan, J.; Liu, S.; Yu, J. Enhanced photovoltaic performance of dye-sensitized solar cells based on TiO_2 nanosheets/graphene composite films. *J. Mater. Chem.* **2012**, *22*, 17027–17036. [CrossRef]

33. Ballesterosgómez, A.; Aragón, A.; Van, D.E.N.; De, B.J.; Covaci, A. Impurities of resorcinol bis(diphenyl phosphate) in plastics and dust collected on electric/electronic material. *Environ. Sci. Technol.* **2016**, *50*, 1934–1940. [CrossRef] [PubMed]

34. Das, O.; Kim, N.K.; Sarmah, A.K.; Bhattacharyya, D. Development of waste based biochar/wool hybrid biocomposites: Flammability characteristics and mechanical properties. *J. Clean. Prod.* **2016**, *144*, 79–89. [CrossRef]

35. Khalili, P.; Tshai, K.Y.; Hui, D.; Kong, I. Synergistic of ammonium polyphosphate and alumina trihydrate as fire retardants for natural fiber reinforced epoxy composite. *Compos. Part B Eng.* **2017**, *14*, 101–110. [CrossRef]

36. Jorge, M.; Fischer, M.; Gomes, J.R.B.; Siquet, C.; Santos, J.C.; Rodrigues, A.E. Accurate model for predicting adsorption of olefins and paraffins on MOFs with open metal sites. *Ind. Eng. Chem. Res.* **2014**, *53*, 15475–15487. [CrossRef]

37. Dai, X.; Cao, Y.; Shi, X.W.; Wang, X.L. Non-isothermal crystallization kinetics, thermal degradation behavior and mechanical properties of poly(lactic acid)/MOF composites prepared by melt-blending methods. *RSC Adv.* **2016**, *6*, 71461–71471. [CrossRef]

38. Han, F.; Liu, Q.; Lai, X.; Li, H.; Zeng, X. Compatibilizing effect of β-cyclodextrin in RDP/phosphorus-containing polyacrylate composite emulsion and its synergism on the flame retardancy of the latex film. *Prog. Org. Coat.* **2014**, *77*, 975–980. [CrossRef]

39. Biswas, B.; Kandola, B.K. The effect of chemically reactive type flame retardant additives on flammability of PES toughened epoxy resin and carbon fiber-reinforced composites. *Polym. Adv. Technol.* **2011**, *22*, 1192–1204. [CrossRef]

40. Mauldin, T.C.; Zammarano, M.; Gilman, J.W.; Shields, J.R.; Boday, D.J. Synthesis and characterization of isosorbide-based polyphosphonates as biobased flame-retardants. *Polym. Chem.* **2014**, *5*, 5139–5146. [CrossRef]

41. Jing, J.; Zhang, Y.; Fang, Z. Diphenolic acid based biphosphate on the properties of polylactic acid: Synthesis, fire behavior and flame retardant mechanism. *Polymer* **2017**, *108*, 29–37. [CrossRef]

42. Abe, H.; Takahashi, N.; Kim, K.J.; Mochizuki, M.; Doi, Y. Effects of residual zinc compounds and chain-end structure on thermal degradation of poly(epsilon-caprolactone). *Biomacromolecules* **2004**, *5*, 1480–1488. [CrossRef] [PubMed]

43. Abe, H.; Takahashi, N.; Kim, K.J.; Mochizuki, M.; Doi, Y. Thermal degradation processes of end-capped poly(L-lactide)s in the presence and absence of residual zinc catalyst. *Biomacromolecules* **2004**, *5*, 1606–1614. [CrossRef] [PubMed]

44. Cui, L.; Tarte, N.H.; Woo, S.I. Effects of modified clay on the morphology and properties of PMMA/clay nanocomposites synthesized by in situ polymerization. *Macromolecules* **2008**, *41*, 4268–4274. [CrossRef]

45. Lee, Y.J.; Huang, J.M.; Kuo, S.W.; Chang, F.C. Low-dielectric, nanoporous polyimide films prepared from PEO–POSS nanoparticles. *Polymer* **2005**, *46*, 10056–10065. [CrossRef]

46. Sahoo, G.; Sarkar, N.; Swain, S.K. The effect of reduced graphene oxide intercalated hybrid organoclay on the dielectric properties of polyvinylidene fluoride nanocomposite films. *Appl. Clay Sci.* **2018**, *162*, 69–82. [CrossRef]

47. Joseph, A.M.; Nagendra, B.; Shaiju, P.; Surendran, K.P.; Gowd, E.B. Aerogels of hierarchically porous syndiotactic polystyrene with a dielectric constant near to air. *J. Mater. Chem. C* **2018**, *6*, 360–368. [CrossRef]

48. Moharana, S.; Mishra, M.K.; Chopkar, M.; Mahaling, R.N. Dielectric properties of three-phase PS-BiFeO$_3$–GNP nanocomposites. *Polym. Bull.* **2017**, *74*, 3707–3719. [CrossRef]

49. Kashi, S.; Gupta, R.K.; Baum, T.; Kao, N.; Bhattacharya, S.N. Dielectric properties and electromagnetic interference shielding effectiveness of graphene-based biodegradable nanocomposites. *Mater. Des.* **2016**, *109*, 68–78. [CrossRef]

50. Babu, R.P.; O'Connor, K.; Seeram, R. Current progress on bio-based polymers and their future trends. *Prog. Biomater.* **2013**, *2*, 8. [CrossRef] [PubMed]

51. Wang, J.Y.; Yang, S.Y.; Huang, Y.L.; Tien, H.W.; Chin, W.K.; Ma, C.C.M. Preparation and properties of graphene oxide/polyimide composite films with low dielectric constant and ultrahigh strength via in situpolymerization. *J. Mater. Chem.* **2011**, *21*, 13569–13575. [CrossRef]

52. Petit, C.; Bandosz, T.J. Engineering the surface of a new class of adsorbents: Metal-organic framework/graphite oxide composites. *J. Colloid Interface Sci.* **2015**, *447*, 139–151. [CrossRef] [PubMed]

53. Mattana, G.; Briand, D.; Marette, A.; Quintero, A.V.; Rooij, N.F.D. Polylactic acid as a biodegradable material for all-solution-processed organic electronic devices. *Org. Electron.* **2015**, *17*, 77–86. [CrossRef]

54. Nelson, J.K.; Fothergill, J.C. Internal charge behaviour of nanocomposites. *Nanotechnology* **2004**, *15*, 586–595. [CrossRef]

55. Lu, J.; Moon, K.S.; Xu, J.; Wong, C.P. Synthesis and dielectric properties of novel high-k polymer composites containing in-situ formed silver nanoparticles for embedded capacitor applications. *J. Mater. Chem.* **2006**, *16*, 1543–1548. [CrossRef]

56. Hu, S.F.; Wong, W.Z.; Liu, S.S.; Wu, Y.C.; Sung, C.L.; Huang, T.Y.; Yang, T.J. A silicon nanowire with a coulomb blockade effect at room temperature. *Adv. Mater.* **2002**, *14*, 736–739. [CrossRef]

57. Tanaka, T.; Kozako, M.; Fuse, N.; Ohki, Y. Proposal of a multi-core model for polymer nanocomposite dielectrics. *IEEE Trans. Dielectr. Electr. Insul.* **2005**, *14*, 669–681. [CrossRef]

 *materials* 

Article

Development of Inherently Flame—Retardant Phosphorylated PLA by Combination of Ring-Opening Polymerization and Reactive Extrusion

Rosica Mincheva [1], Hazar Guemiza [2], Chaimaa Hidan [2], Sébastien Moins [1], Olivier Coulembier [1], Philippe Dubois [1,2] and Fouad Laoutid [2,*]

[1] Laboratory of Polymeric and Composite Materials, University of Mons, Place du Parc 23, 7000 Mons, Belgium; rosica.mincheva@umons.ac.be (R.M.); Sebastien.MOINS@umons.ac.be (S.M.); Olivier.coulembier@umons.ac.be (O.C.); philippe.dubois@umons.ac.be (P.D.)

[2] Polymeric and Composite Materials Unit, Materia Nova Research Center, Nicolas Copernic 3, 7000 Mons, Belgium; hazarguemiza@gmail.com (H.G.); chai.hidan@gmail.com (C.H.)

* Correspondence: fouad.laoutid@materianova.be; Tel.: +32-(0)65-55-49-78

Received: 14 November 2019; Accepted: 11 December 2019; Published: 18 December 2019

Abstract: In this study, a highly efficient flame-retardant bioplastic poly(lactide) was developed by covalently incorporating flame-retardant DOPO, that is, 9,10-dihydro-oxa-10-phosphaphenanthrene-10-oxide. To that end, a three-step strategy that combines the catalyzed ring-opening polymerization (ROP) of L,L-lactide (L,L-LA) in bulk from a pre-synthesized DOPO-diamine initiator, followed by bulk chain-coupling reaction by reactive extrusion of the so-obtained phosphorylated polylactide (PLA) oligomers (DOPO-PLA) with hexamethylene diisocyanate (HDI), is described. The flame retardancy of the phosphorylated PLA (DOPO-PLA-PU) was investigated by mass loss cone calorimetry and UL-94 tests. As compared with a commercially available PLA matrix, phosphorylated PLA shows superior flame-retardant properties, that is, (i) significant reduction of both the peak of heat release rate (pHRR) and total heat release (THR) by 35% and 36%, respectively, and (ii) V0 classification at UL-94 test. Comparisons between simple physical DOPO-diamine/PLA blends and a DOPO-PLA-PU material were also performed. The results evidenced the superior flame-retardant behavior of phosphorylated PLA obtained by a reactive pathway.

Keywords: reactive flame retardancy; PLA ROP; chain extension; DOPO

---

1. Introduction

Bio-based polymers have recently attracted a growing interest to improve the sustainability of plastics. These green materials gained an increasing demand for durable applications because they present a reduced carbon footprint compared with those produced from fossil carbon. However, their use in high-performance applications is limited because of their high flammability. In fact, fire retardancy of polymeric materials has become an important requirement prior to being used in several technical applications.

Polylactide (PLA) is one of the most promising biopolymers that is increasingly being used for technical applications [1,2], owing to its various advantages. It is produced from annually renewable resources and it presents a high stiffness, high degree of transparency in addition to its relatively low cost, and large production volume. However, PLA suffers from some shortcomings because some of its properties such as impact resistance, ductility, tensile strength, service temperature, long-term stability, and flame-retardant behavior need to be improved.

Currently, the additive way is the most common approach for enhancing PLA flame-retardant properties. It consists of the physical blending of PLA with flame-retardant (FR) additives such as organic phosphorus [3–5], intumescent systems [6,7], and mineral fillers [8].

Among these, the organic phosphorus flame retardants are the most adapted for PLA. The presence of phosphorous promotes the formation of crosslinked or carbonized structures that act as an insulator layer during the combustion, thus protecting the polymer underneath from the heat and limiting the volatilization of fuel [9]. The association of phosphorus-based flame-retardant additives with bio-based char forming agents such as lignin [10–13], cellulose [14,15], or tannic acid [16,17] presents an efficient way for enhancing flame retardant performances and reducing the environmental impact of the FR system. An incorporation rate of at least 20 wt.% is required prior to obtaining enhanced flame-retardant performances. However, PLA is very sensitive to thermal degradation during melt processing and the incorporation of these additives generally induces an important reduction of polymer molecular weight, significantly affecting its mechanical properties.

The integration of phosphorous directly along the PLA chain presents an interesting alternative to the additive way. Using a reactive phosphorous-based molecule as a PLA chain extension is an easy way allowing the incorporation of phosphorous in the backbone of the PLA macromolecule. With this view, Wang et al. [18] used ethyl phosphorodichloridate for chain-extending dihydroxyl terminated pre-PLA. The resulting materials (PPLA) showed relevant flame-retardant properties and the use of only 5 wt.% PPLA into PLA enables obtaining a blend with good flame-retardant properties. Up to 10 wt.%, the so-obtained material presented Limited Oxygen Index of 35, a lower peak heat release rate (pHRR), a longer time to ignition, and V0 classification at the UL-94 test.

In this work, a three-step strategy aiming to prepare novel inherently flame-retardant polylactide-based materials is described. First, a bifunctional phosphorous-based diamine was prepared through the reaction between 9,10-dihydro-oxa-10-phosphaphenanthrene-10-oxide (DOPO) and 4,4′-diaminobenzophenone (DABP). This bifunctional diamine was thus used as an initiator for bulk ring-opening polymerization (ROP) of L,L-lactide (L,L-LA). The last step consisted of the bulk chain-coupling reaction of the so-obtained phosphorylated PLA oligomers with hexamethylene diisocyanate (HDI). DOPO was selected for this application because of its flame-retardant action, which is effective at a low P content (1 to 2 wt.%). In fact, DOPO presents an interesting flame-retardant effect because it acts in both condensed and gas phases [19]. During pyrolysis, DOPO-based flame retardants release reactive species in the gas phase (most probably PO°) that are very effective to achieve efficient flame inhibition, while the phosphorous remaining in the condensed phase promotes the formation of char residue.

The structure of the obtained phosphorylated PLA oligomers (DOPO-PLA) as well as the corresponding chain extended material (DOPO-PLA-PU) were characterized using Proton and Phosphorus nuclear magnetic resonance (1H & 31P NMR respectively) as well as by size exclusion chromatography (SEC), while their thermal properties and flame-retardant properties were evaluated by differential scanning calorimetry (DSC), thermal gravimetrical analysis (TGA), UL-94, and cone calorimeter tests, respectively. The obtained results evidenced superior flame-retardant properties of the novel phosphorylated PLA when tested alone and in blend with commercial PLA.

2. Materials and Methods

2.1. Materials

The L,L-LA (Purasorb®L, optical purity > 99.5%, MW = 144 g·mol^{-1}, free acid < 1 meq·kg^{-1}, water content < 0.02%, Purac Biochem BV, The Netherlands) and hexamethylene diisocyanate (HDI, purris. ≥99.0%, MW = 168 g·mol^{-1}, Sigma-Aldrich, (Diegem, Belgium) were stored in a glove box prior to use. Tin(II) 2-ethylhexanoate (Sn(Oct)$_2$, 95%, MW = 405 g·mol^{-1}) and 4,4′-diaminobenzophenone (DABP, 97%, MW = 212 g·mol^{-1}) from Sigma-Aldrich (now Merck) were used as received. 9,10-Dihydro-oxa-10-phosphaphenanthrene-10-oxide (DOPO, >97 %, MW = 216 g·mol^{-1}, ADD APT

Chemicals AG, The Netherlands) was recrystallized from tetrahydrofurane (THF) prior to use. All solvents (that is, methanol, chloroform, and THF; VWR, Belgium) were of analytical grade and used as received. Polylactide (Ingeo™ Biopolymer 3052D) from NatureWorks (Minnetonka, MN, USA) was used in this study as a commercial PLA resin. PLA 3052D is a semi-crystalline polymer with an average molecular weight Mn of 90,000 g/mol.

2.2. Syntheses

DOPO-diamine

The DOPO-diamine was obtained according to the literature [20]. Briefly, 32 g (0.15 mol, 1 eq) of DOPO and 5.35 g (0.25 mol, 1.7 eq) of DABP were placed into a 250 ml round-bottom glass-flask and heated at 180 °C for 3 h. The obtained homogeneous mixture was then allowed to cool-down to 100 °C, followed by the addition of 150 mL of toluene in order to precipitate the DOPO-diamine. The product was filtered and recrystallized repeatedly (×3) from THF (150 mL). Yield: 55%. 1H NMR (500 MHz, DMSO-d6, δ ppm): 8.5–6.5 (m, 20H, CHphenyl), 5.9 (br.dt, 4H, CHphenyl-NH2), 4.9 (br.s, 4H, NH2). 31P NMR (500 MHz, DMSO-d6, δ ppm): 31.78, 30.12. DSC (H/C/H, 10 °C·min^{-1}, 0–200 °C): mp. 325 °C.

DOPO-PLA

DOPO-PLA was prepared via DOPO-diamine initiated ROP of L,L-LA in bulk following the procedure described hereafter: ROP of L, L-LA as initiated by DOPO diamine was performed in specially designed 250 mL Inox™ reactors (Maximator France, Rantigny, France) at 180 °C and a stirring rate of 50 rpm for 90 min. Slight nitrogen pressure (0.4 bar) was introduced in order to ensure an inert atmosphere and a premixing time of 10 min was used. Then, 50 g of L,L-LA (347 mmol) was mixed with 22 g of DOPO-diamine (35 mmol) and 1.4 g (3 mmol) of Sn(Oct)$_2$. The molar ratio [L,L-LA]0/[DOPO-diamine]0 was fixed to 10. Stannous (II) octoate was used as catalyst at molar ratio [L,L-LA]0/[Sn(Oct)$_2$]0 = 100.

The as-obtained DOPO-PLA was recovered by solubilization in THF and precipitation in cold methanol.

DOPO-PLA PUs

DOPO-PLA PU was prepared via reactive extrusion using a 15 cm^3 vertical corotational twin-screw DSM microcompounder (DSM, Sittard, The Netherlands), equipped with nitrogen inlet and water-cooling systems at 160 °C. Typically, 14 g of DOPO-PLA (5 mmol) was introduced into the extruder at 30 rpm and allowed to melt for 3 min, after which 0.98 mL HDI (5.8 mmol) was added and the reaction was allowed to continue at 30 rpm for 90 min.

Blending

Blending of DOPO-PLA-PU with commercial PLA was carried out in a Brabender internal mixer at 180 °C for 10 min (2 min mixing at 30 rpm followed by 8 min at 60 rpm). For the mass loss calorimeter test, plates (100 × 100 × 3 mm^3) were compression molded at 180 °C using an Agila PE20 hydraulic press. More precisely, the material was first pressed at a low pressure for 200 s (three degassing cycles), followed by a high-pressure cycle at 150 bars for 180 s. The samples were then cooled down under pressure (80 bars).

2.3. Methods

1H and 31P NMR spectra were collected with a Bruker AMX-500 at a frequency of 500 MHz instruments, respectively, in Hexadeuterodimethyl sulfoxide (DMSO-d6).

Size exclusion chromatography (SEC) was used to determine the molar mass of the prepared materials. Analyses were performed at 30 °C in chloroform (CHCl$_3$) using an Agilent liquid chromatograph equipped with an Agilent degasser, an isocratic High-performance liquid chromatography (HPLC) pump (flow rate = 1 mL·min^{-1}), an Agilent autosampler (loop volume = 200 μL, solution conc. = 2.5 mg·mL^{-1}), an Agilent-DRI refractive index detector, and three columns: a PL gel 10 μm guard column and two PL gel Mixed-D 10 μm columns (linear columns for separation of MWPS ranging from 500 to 106 g·mol^{-1}). Polystyrene standards were used for calibration.

Thermal gravimetrical analyses (TGAs) were performed with a TA Q500 thermogravimetric analyzer (Zellik, Belgium). The weight loss was recorded upon heating the samples (having initial masses of ca. 10 mg) at 20 °C·min^{-1} from 30 to 700 °C, under N_2 flow of 60 mL·min^{-1}. The referred temperatures of maximum degradation rate (Tmax) were taken as the inflection point of the sigmoidal steps (i.e., the maxima of the first-derivative curve).

Differential scanning calorimetry (DSC) measurements were performed with a TA Q200 calorimeter (Zellik, Belgium) calibrated with high purity indium and operating under N_2 flow. The sample weight was about 5 mg and a scanning rate of 10 °C·min^{-1} was employed in all the runs. Each run followed a heat/cool/heat (H/C/H) procedure from 0 to 200 °C. The reported thermal transition values, Tg (glass transition temperature) and Tm (melting temperature), were defined as the midpoints of the sigmoidal curve, minima of the exotherms, and maxima of the endotherms, respectively.

The fire behavior of the DOPO-PLA-PU was evaluated using the mass loss cone calorimeter test at 35 kW·m^{-2}. Cone calorimeter is one of the most used devices to assess the flammability of materials at bench scale. The peak of heat release rate (pHRR), time to ignition (TTI), as well as total heat release (THR), which are considered as the most important parameters in this fire test, were considered. A high pHRR and a low TTI may cause both fast ignition and rapid-fire propagation. Mass loss cone calorimeter tests were performed according to ISO 13927 standard procedures with a Fire Testing Technology Limited (FTT) mass loss cone calorimeter. Samples (100 × 100 × 3 mm^3) were exposed to an external heat flux of 35 kW·m^{-2}, corresponding to common heat flux in a mild fire scenario. UL-94 test was performed on films of 0.8 mm thickness.

3. Results

3.1. DOPO-Diamine Initiator Preparation and Characterizations

DOPO-diamine was synthesized through a well-established procedure (Scheme 1A) starting from 9,10-dihydro-oxa-10-phosphaphenanthrene-10-oxide (DOPO) and 4,4'-diaminobenzophenone (DABP). [20] Global yield reached 55% after recovery of the product obtained after three hours at 180 °C. In accordance to the state-of-the-art, 1H and 31P NMR characterizations concluded on the formation of the expected DOPO-diamine.

Prior to its use as a polymerization initiator, the thermal stability of the DOPO-diamine was studied by thermogravimetric analysis (TGA) and compared to both DOPO and DABP reactants (Figure 1). Clearly, the DOPO-diamine is characterized by a well-higher thermal stability as compared with its precursors. With a thermal degradation starting above 300 °C, the DOPO-diamine TGA analysis also reveals the formation of a char residue of 27 wt.% at circa 450 °C. Those results are of prime importance and indicate that DOPO-diamine could be processed at PLA melt processing (180 °C) without undergoing any thermal degradation.

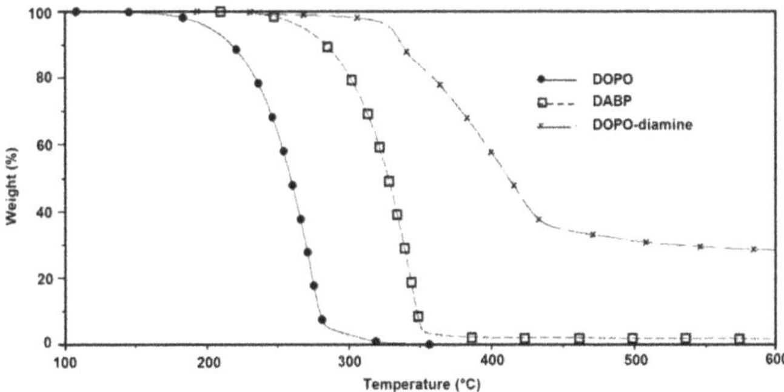

Scheme 1. Three-step synthetic pathway to 9,10-dihydro-oxa-10-phosphaphenanthrene-10-oxide (DOPO)-polylactide (PLA) PUs: (**A**) Synthesis of DOPO-diamine; (**B**) DOPO-initiated bulk ring-opening polymerization (ROP) of L,L-lactide (L,L-LA); (**C**) chain coupling reaction.

Figure 1. Thermogravimetric analyses (TGAs) performed on 9,10-dihydro-oxa-10-phosphaphenanthrene-10-oxide (DOPO), DABP, and DOPO-diamine at 20 °C/min under nitrogen.

3.2. Synthesis and Properties of DOPO-PLA Oligomers

The DOPO-diamine was then employed as a telechelic initiator for the ring-opening polymerization (ROP) of L,L-LA in the presence of tin(II) octanoate (Sn(Oct)$_2$). Duda et al. showed that the mechanism of LA ROP from primary amine and catalyzed by Sn(Oct)$_2$ does not differ from an ROP process initiated from an alcohol exogenous initiator [21]. In the specific case of the DOPO-diamine, a controlled "coordination-insertion" ROP of L,L-LA would proceed from both primary amines, leading to a DOPO core functionalized by two PLA segments end-capped by hydroxyl groups (Scheme 1B). To get rid of the inherent insolubility of the DOP-diamine in most of the studied organic solvents, the L,L-LA ROP process was performed in bulk at 180 °C for an initial monomer-to-initiator ratio ([L,L-LA]0/[DOPO-diamine]0) of 10. Note here that such a ratio was selected to produce an oligo-PLA composed by a sufficiently high atomic percentage in phosphorus atoms (4 wt.%) to impact the fire properties. To ensure a controlled process, a [DOPO-diamine]0/[Sn(Oct)$_2$]0 molar ratio of 12/1 was used [22]. After 90 min of reaction, a 1H NMR analysis of the crude medium revealed a conversion in L,L-LA of 96%. By comparing the relative intensities of free DOPO-diamine to the one incorporated into the DOPO-PLA, an efficacy of initiation as high as 70 mol% was calculated. After precipitation of the sample, and considering that the DOPO-diamine initiated the L,L-LA ROP symmetrically, an experimental molar mass (Mn, NMR) of 2100 g/mol was calculated by 1H NMR spectroscopy, which is in good agreement with the theoretical value (Mn, th = 2000 g/mol).

In comparison with the two phosphorus signals observed by analysis of the pristine DOPO-diamine before polymerization (at 29 and 31 ppm), the 31P NMR analysis of the as-obtained DOPO-PLA confirmed the incorporation of the DOPO initiator into the PLA chain by the presence of one single phosphorus peak at 30.5 ppm (Figure 2) [20].

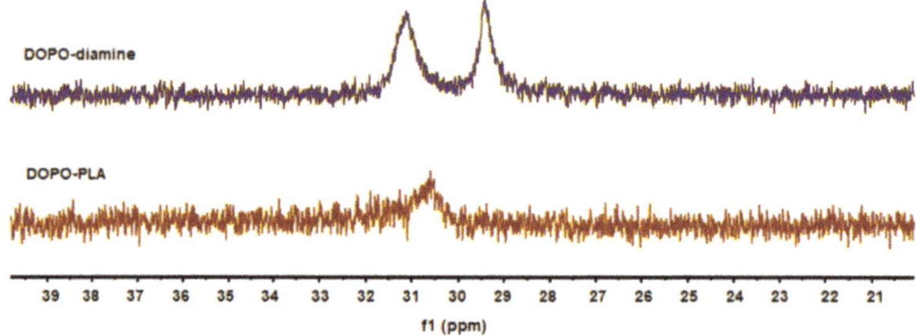

Figure 2. ^{31}P NMR analyses (zoomed between 40 and 20 ppm) of DOPO-diamine and DOPO-polylactide (PLA)-PU.

Size exclusion chromatography (SEC) traces of the polylactide generated from the DOPO-diamine initiator (Mn = 2700 g·mol^{-1}, Đ = 1.5) and using both refractive index and UV detectors (254 nm, respectively) clearly show that DOPO is distributed throughout the sample (Figure 3). Indeed, SEC in dual RI and UV detection is a well-known method providing information on the composition at any point in the molar mass distribution when at least one component absorbs at a suitable UV-wavelength [23]. With this respect and knowing that DOPO is the only element in DOPO-PLA that absorbs at a UV-wavelength, it was possible to confirm its presence at any point of the molar mass distribution by the UV elution curve precisely following the RI one, as shown in Figure 3.

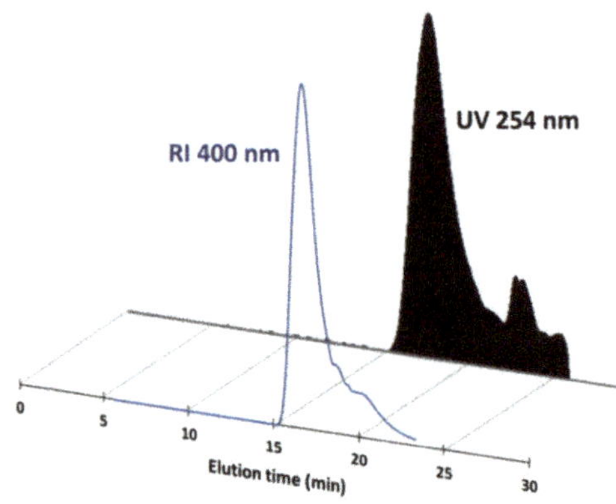

Figure 3. Size exclusion chromatography (SEC) elution curves of DOPO-PLA using both RI and UV detectors.

Finally, a TGA analysis was performed on the DOPO-PLA oligomer (Figure 4) and the results summarized in Table 1. In comparison with high commercial molecular weight PLA (see experimental section), DOPO-PLA oligomers show lower thermal stability because their decomposition starts at 241 °C ($T_{5\%}$), while the weight loss of commercial PLA occurs at 300 °C. The different thermal stability observed between these two materials is the result of their high difference in molecular weight. PLA thermal stability is highly dependent on Mn and low molecular weight chain starts to decompose at a lower temperature [24–26]. Interestingly, TGA curves also evidence another behavior. In fact, during their thermal decomposition, DOPO-PLA oligomers generate some char, while commercial PLA totally degrades during the analysis. This result highlights the ability of DOPO to act in the condensed phase by promoting the formation of char structures.

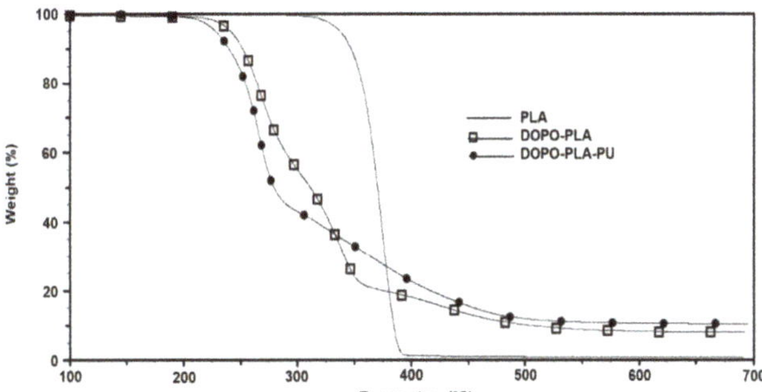

Figure 4. TGA curves of phosphorylated PLA oligomers (DOPO-PLA) and chain extended phosphorylated oligomers (DOPO-PLA-PU) in comparison with commercial PLA, 20 °C/min under nitrogen.

Table 1. Characteristic degradation temperatures and residues amounts from thermogravimetric analysis (TGA) experiments (under N_2 and at 20 °C/min) for commercial polylactide (PLA), phosphorylated PLA oligomers (9,10-dihydro-oxa-10-phosphaphenanthrene-10-oxide (DOPO)-PLA), chain extended phosphorylated PLA oligomers (DOPO-PLA-PU), as well as PLA/DOPO-PLA-PU blend.

Sample	$T_{5\%}$ (°C)	T_{max} (°C)	Residue at 700 °C (%)
PLA	337	374	0.4
DOPO-PLA	241	267; 338	8.5
DOPO-PLA-PU	228	265	10.4
50% DOPO-PLA-PU/50% PLA	218	255	5

3.3. Synthesis and Properties of DOPO-PLA-PU by Chain Extension

DOPO-PLA oligomers were thus further subjected to chain-extension reaction using HDI via reactive extrusion in order to increase their overall molecular weight. As evidenced by 1H NMR spectroscopy (Figure 5, inset), DOPO-PLA reacted well with the diisocyanate moiety. Indeed, the signal corresponding to the hydroxymethine-PLA end groups (–CH(CH$_3$)OH), initially present at 4.25 ppm (Figure 5), almost completely disappeared, as well as the hydroxy proton end groups (-CH(CH3)OH) at 5.48 ppm, to the benefit of new urethane bonds. While integrations of amide protons (Hc, 10 ppm) and methine repeating units (Ha, 5.18 ppm) are logically identical before and after the chain extension experiment, the diminushion of the methine PLA end groups (4.25 ppm) from an integration of 1.54 to 0.39 allows an efficiency of reaction of 78% to be calculated. Note here that, owing to the limited PUs solubility in common solvents [27] and the possible interactions of the as-obtained DOPO-PLA-PU with the SEC columns, resulting in truncated Mn value, no SEC analysis was realized. It is worth mentioning that the incorporation of urethane functions in the backbone of PLA macromolecule does not affect its compostability [28].

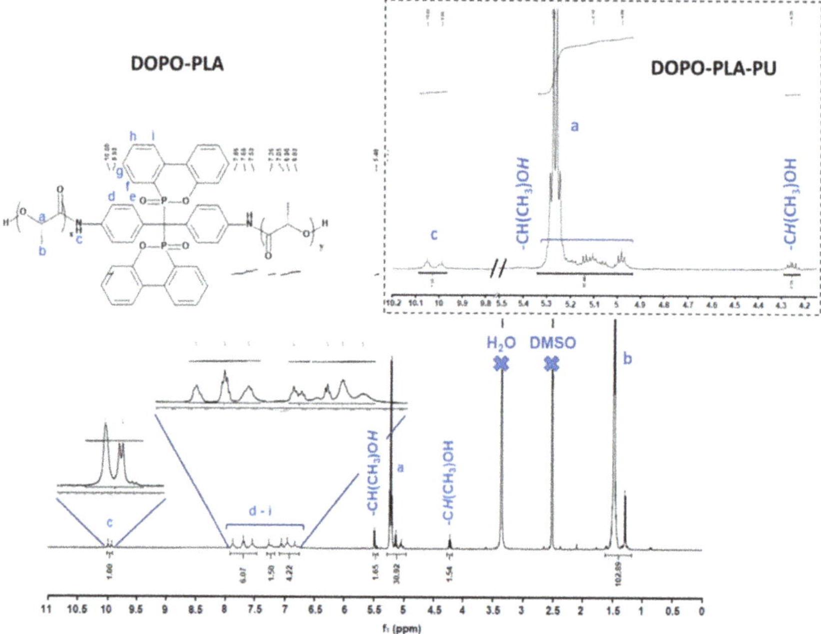

Figure 5. ^{1}H NMR spectrum of DOPO-PLA recorded in DMSO-d6. The figure inset shows the ^{1}H NMR analysis of DOPO-PLA-PU (only the zoom between δ = 4.2 and 10.2 ppm is shown for clarity).

The resulting DOPO-PLA-PU produced is in the form of a transparent yellow material (Figure 6). The DSC results specify that it is an amorphous polymer characterized by a Tg of 60 °C, similar to that of unmodified PLA (Figure 7 and Table 2). The presence of DOPO-diamine and urethane functions hinders the crystallization of DOPO-PLA-PU polymer. However, the thermal stability of chain extended DOPO-PLA is slightly reduced in comparison with the corresponding oligomers. Further, the presence of urethane moieties also enables a slight increase in the amount of residue (+2%) (Figure 4 and Table 1).

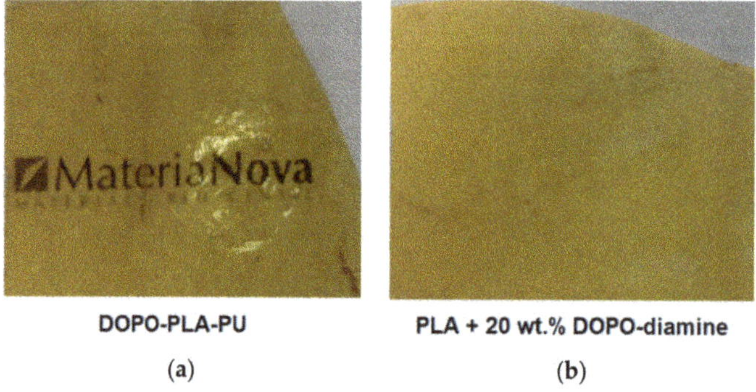

Figure 6. Films (thickness of ca. 0.8 mm) of transparent DOPO-PLA-PU (**a**) and the opaque blend of commercial PLA containing 20 wt.% of DOPO-diamine as additive (**b**).

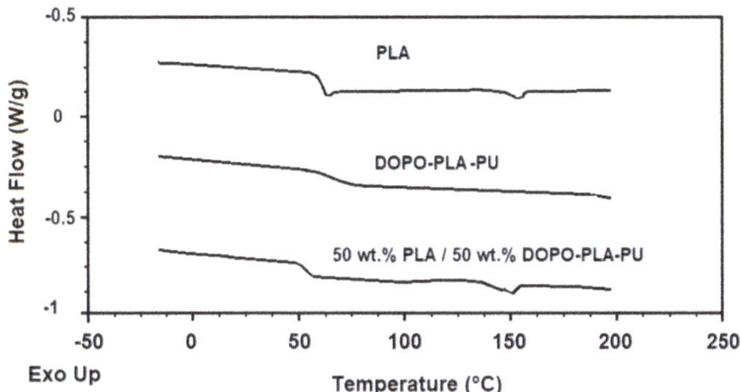

Figure 7. Differential scanning calorimetry (DSC) heating curves (second run) of commercially available PLA, DOPO-PLA-PU, and their blend 50/50 (wt/wt.%).

Table 2. Glass transition (Tg), melting temperature (Tm), and melting enthalpies (ΔHm) obtained by differential scanning calorimetry (DSC) second heating run.

Sample	T_g (°C)	T_m (°C)	ΔH_m (J/g)
PLA	60	154	2
DOPO-PLA-PU	60.5	–	–
50% DOPO-PLA-PU/50% PLA	50.5	150	3

The flame-retardant behavior of DOPO-PLA-PU was assessed using mass loss cone calorimeter. For comparison, Figure 8 and Table 3 summarize the results obtained during the combustion of DOPO-PLA-PU and the commercially available PLA. Pristine PLA exhibits a strong combustion, starting after 40 s, consuming all the material and releasing a total heat of about 54 MJ/m^2 with a pHRR of 580 kW/m^2. In the case of DOPO-PLA-PU, a lower pHRR (−35%) and THR (−36%), as well as a reduction of the time to ignition from 40 to 16 s, were observed. It is worth mentioning that, during the combustion of DOPO-PLA-PU, a char residue was formed at the surface of the burning material, evidencing the condensed phase action of the DOPO contained in the material. However, this char was not thermally stable and only a few pieces remain at the end of the test. The formation of the char during the combustion is responsible for the reduction of HRR, but it is not excluded that a gas phase may also occur. Indeed, DOPO and its derivatives act in both gas and condensed phases by the release of low molecular weight phosphorus based species that can scavenge the H° and OH° radicals in the flame and the generation of char layer. [29] The flame-retardant effect in the gas phase cannot be excluded as UL-94 tests performed on films of 0.8 mm thickness highlighted that the material was very difficult to ignite. A flame inhibition owing to the generation of active P-based species seems to be responsible for this resistance to ignition in addition to the char formation. In fact, during the UL-94 test, the material is very difficult to ignite, and melts, but drops do not induce any cotton ignition and only a black residue is formed at the surface of cotton (Figure 9). The low time to ignition observed in the case of DOPO-PLA-PU is the result of the low thermal stability of this polymer that starts to decompose around 100 °C, thus earlier than the comparative commercial PLA (Figure 4). The reduced thermal stability of DOPO-PLA-PU is likely the result of the lower molecular weight of DOPO-PLA-PU with respect to the commercial PLA. In fact, Figure 4 showed that the TGA curve of DOPO-PLA-PU is close to that of the starting phosphorylated PLA oligomers, meaning the urethane functions do not affect the material thermal stability.

The high weight phosphorus content in DOPO-PLA-PU (4%) is thus beneficial and allows obtaining a material presenting good flame-retardant performances. However, an in-depth study is needed for better establishing the real contribution of DOPO when combined with DABP and integrated in the backbone of PLA macromolecules.

Interestingly enough, when DOPO-diamine is physically added to PLA at 10 wt.% and 20 wt.% content, the so-obtained material does not present any flame-retardant behavior during the cone calorimeter test (Figure 8 and Table 3). Moreover, using the additive route also affects the transparency of the material, which becomes opaque (Figure 6b).

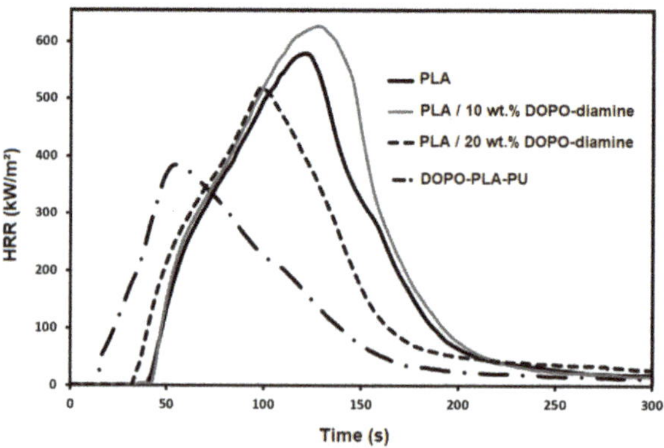

Figure 8. Heat release rate (HRR) curves obtained during the mass loss cone calorimeter test at 35 kW/m^2.

Table 3. Cone calorimeter (35 kW/m^2) and UL-94 results. pHRR, peak heat release rate; TTI, time to ignition; THR, total heat release.

Sample	TTI (s)	pHRR (kW/m^2)	pHRR Reduction (%)	THR (MJ/m^2)	THR Reduction (%)	UL-94 Classification
PLA	40	580	–	54	–	No rating
PLA/10 wt.% DOPO-diamine	42	620	No reduction	61.7	No reduction	No rating
PLA/20 wt.% DOPO-diamine	34	520	No reduction	51	No reduction	No rating
DOPO-PLA-PU	16	380	35	34.5	–36	V0

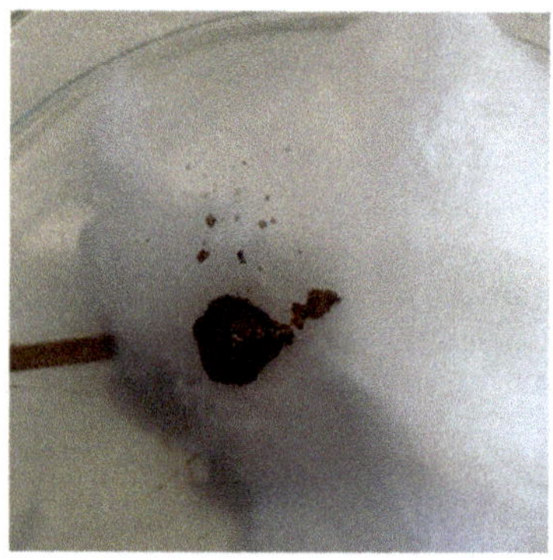

Figure 9. Photograph showing the formation of black residue at the surface of cotton after the UL-94 test.

Melt blending DOPO-PLA-PU with commercial PLA presents another interesting way for taking advantage of the good flame-retardant properties of DOPO-PLA-PU. With this view, a new blend was prepared by melt blending the two polymers in an internal mixer. TGA analysis (Figure 10) shows that the incorporation of 50 wt.% DOPO-PLA-PU induces an important reduction of PLA thermal stability, which becomes similar to that of the phosphorylated PLA-PU, but unfortunately does not promote the formation of further residue, as all PLAs continue to decompose and only a final residue corresponding to that generated by the thermal degradation of DOPO-PLA-PU is obtained. Interestingly, DSC analyses show that the presence of DOPO-PLA-PU also induces a reduction of Tg of the blend that decreases from 60 °C for both polymers separately to 50 °C when combined. This result evidences the good miscibility between both polymers where DOPO-PLA-PU acts as plasticizer for the PLA phase. Moreover, the so-obtained material enables obtaining transparent films.

Blending DOPO-PLA-PU (50 wt.%) with commercial PLA presents a good strategy because the UL-94 test, performed on 0.8 mm thick films, allows for reaching a remarkable V0 classification, while commercial PLA simply burns after the first flame application, forming burning drops that induce cotton inflammation. This strategy could be also applied for PMMA, as PLA and PMMA demonstrate good miscibility [30] and blending DOPO-PLA-PU with pristine PMMA could enable enhancing PMMA flame-retardant properties, while maintaining the material transparency.

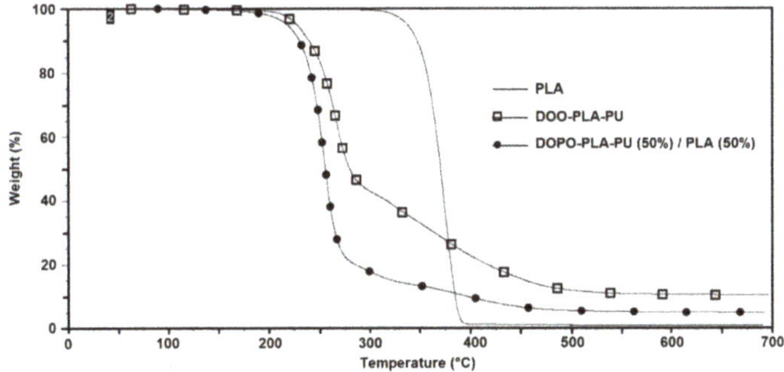

Figure 10. TGA curves under nitrogen (20 °C/min).

4. Conclusions

Organophosphorus flame-retardant, a bifunctional DOPO-diamine, was successfully prepared and chemically incorporated into the PLA backbone by combining ring-opening polymerization and reactive extrusion. The bifunctional amine was first prepared by reacting DOPO with 4,4'-diaminobenzophenone and used as an initiator for bulk ring-opening polymerization of L,L-lactide (L,L-LA). The last step consisted of the bulk chain-coupling reaction of the so-obtained phosphorylated PLA oligomers with hexamethylene diisocyanate (HDI) by reactive extrusion.

Chemical structures of the phosphorylated PLA oligomers and its chain extended homologue were confirmed by 1H, 31P NMR, and SEC analyses. The thermal analysis revealed that DOPO-PLA oligomers presented lower thermal stability than commercially available PLA, but superior flame-retardant properties. Cone calorimeter tests evidenced a pHRR and THR reduction of about 35 and 36%, respectively, for DOPO-PLA-PU with respect to commercially available PLA. Moreover, DOPO-PLA-PU reached V0 classification at the UL-94 test. Combining DOPO-PLA-PU with PLA was shown to present an interesting way for developing transparent PLA films (0.8 mm thickness) with V0 classification.

All these results demonstrated that this novel synthesis route enables the development of inherent flame-retardant PLA. Ongoing works are underway in our laboratory to understand the mode of action of DOPO-diamine when incorporated into PLA backbone.

Author Contributions: Conceptualization, R.M., O.C., P.D. and F.L.; methodology, R.M. and F.L.; validation, R.M. and F.L.; formal analysis, R.M. and F.L.; investigation, H.G., C.H. and S.M.; resources, R.M, O.C. and F.L.; writing—original draft preparation, R.M. and F.L.; writing—review and editing, O.C., F.L. and P.D.; visualization, R.M. and F.L.; supervision, P.D. and F.L.; project administration, F.L.; funding acquisition, P.D. All authors have read and agreed to the published version of the manuscript.

Funding: This research was funded by Wallonia and the European Commission "FSE and FEDER" in the frame of Phasing-out Hainaut and the LCFM-BIOMASS and MACOBIO projects, and the National Fund for Scientific Research (F.R.S.-FNRS).

Conflicts of Interest: The authors declare no conflict of interest.

References

1. Kuczynski, J.; Boday, D.J. Bio-based materials for high-end electronics applications. *Int. J. Sustain. Dev. World* **2012**, *19*, 557–563. [CrossRef]
2. Wertz, J.T.; Mauldin, T.C.; Boday, D.J. Polylactic acid with improved heat deflection temperatures and self-healing properties for durable goods applications. *Appl. Mater. Interfaces* **2014**, *6*, 18511–18516. [CrossRef] [PubMed]
3. Costes, L.; Laoutid, F.; Dumazert, L.; Lopez-cuesta, J.-M.; Brohez, S.; Delvosalle, C.; Dubois, P. Metallic phytates as efficient bio-based phosphorous flame retardant additives for poly (lactic acid). *Polym. Degrad. Stab.* **2015**, *119*, 217–227. [CrossRef]
4. Wei, L.-L.; Wang, D.-Y.; Chen, H.-B.; Chen, L.; Wang, X.-L.; Wang, Y.-Z. Effect of a phosphorus-containing flame retardant on the thermal properties and ease of ignition of poly (lactic acid). *Polym. Degrad. Stab.* **2011**, *96*, 1557–1561. [CrossRef]
5. Tang, G.; Wang, X.; Zhang, R.; Wang, B.B.; Hong, N.N.; Hu, Y.; Song, L.; Gong, X. Effect of rare earth hypophosphite salts on the fire performance of biobased polylactide composites. *Ind. Eng. Chem. Res.* **2013**, *52*, 7362–7372. [CrossRef]
6. Cayla, A.; Rault, F.; Giraud, S.; Salaün, F.; Fierro, V.; Celzard, A. PLA with Intumescent System Containing Lignin and Ammonium Polyphosphate for Flame Retardant Textile. *Polymers* **2016**, *8*, 331. [CrossRef]
7. Réti, C.; Casetta, M.; Duquesne, S.; Bourbigot, S.; Delobel, R. Flammability properties of intumescent PLA including starch and lignin. *Polym. Adv. Technol.* **2008**, *19*, 628–635. [CrossRef]
8. Murariu, M.; Bonnaud, L.; Paint, Y.; Fontaine, G.; Bourbigot, S.; Dubois, P. New trends in polylactide (PLA)-based materials: "Green" PLA-Calcium sulfate (nano)composites tailored with flame retardant properties. *Polym. Degrad. Stab.* **2010**, *95*, 374–381. [CrossRef]
9. Laoutid, F.; Bonnaud, L.; Alexandre, M.; Lopez-Cuesta, J.-M.; Dubois, P. New prospects in flame retardant polymer materials: From fundamentals to nanocomposites. *Mater. Sci. Eng. R Rep.* **2009**, *63*, 100–125. [CrossRef]
10. Costes, L.; Laoutid, F.; Aguedo, M.; Richel, A.; Brohez, S.; Delvosalle, C.; Dubois, P. Phosphorus and nitrogen derivatization as efficient route for improvement of lignin flame retardant action in PLA. *Eur. Polym. J.* **2016**, *84*, 652–667. [CrossRef]
11. Costes, L.; Laoutid, F.; Brohez, S.; Delvosalle, C.; Dubois, P. Phytic acid–lignin combination: A simple and efficient route for enhancing thermal and flame-retardant properties of polylactide. *Eur. Polym. J.* **2017**, *94*, 270–285. [CrossRef]
12. Prieur, B.; Meub, M.; Wittemann, M.; Klein, R.; Bellayer, S.; Fontaine, G.; Bourbigot, S. Phosphorylation of lignin to flame retard acrylonitrile butadiene styrene (ABS). *Polym. Degrad. Stab.* **2016**, *127*, 32–43. [CrossRef]
13. Chollet, B.; Lopez-Cuesta, J.M.; Laoutid, F.; Ferry, L. Lignin Nanoparticles as a Promising Way for Enhancing Lignin Flame Retardant Effect in Polylactide. *Materials* **2019**, *12*, 2132. [CrossRef] [PubMed]
14. Costes, L.; Laoutid, F.; Khelifa, F.; Rose, G.; Brohez, S.; Delvosalle, C.; Dubois, P. Cellulose/phosphorus combinations for sustainable fire retarded polylactide. *Eur. Polym. J.* **2016**, *74*, 218–228. [CrossRef]
15. Fox, D.M.; Novy, M.; Brown, K.; Zammarano, M.; Harris, R.H., Jr.; Murariu, M.; Carthy, E.D.M.; Seppala, J.E.; Gilman, W.J. Flame retarded poly (lactic acid) using POSS-modified cellulose. 2. Effects of intumescing flame retardant formulations on polymer degradation and composite physical properties. *Polym. Degrad. Stab.* **2014**, *106*, 54–62. [CrossRef]
16. Laoutid, F.; Vahabi, H.; Shabanian, M.; Aryanasab, F.; Zarrintaj, P.; Saeb, M.R. A new direction in design of bio-based flame retardants for poly (lactic acid). *Fire Mater.* **2018**, *42*, 914–924. [CrossRef]

17. Laoutid, F.; Karaseva, V.; Costes, L.; Brohez, S.; Mincheva, R.; Dubois, P. Novel bio-based flame retardant systems derived from tannic acid. *J. Renew. Mater.* **2018**, *6*, 559–572. [CrossRef]

18. De-Yi, W.; Yan-Peng, S.; Ling, L.; Xiu-Li, W.; Yu-Zhong, W. A novel phosphorus-containing poly (lactic acid) toward its flame retardation. *Polymer* **2011**, *52*, 233–238.

19. Schartel, B. Phosphorus-based flame retardancy mechanisms—old hat or a starting point for future development? *Materials* **2010**, *3*, 4710–4745. [CrossRef]

20. Wu, C.S.; Liu, Y.L.; Chiu, Y.-S. Synthesis and characterization of new organosoluble polyaspartimides containing phosphorous. *Polymer* **2002**, *43*, 1773–1779. [CrossRef]

21. Kowalski, A.; Libiszowski, J.; Biela, T.; Cypryk, M.; Duba, A.; Penczek, S. Kinetics and Mechanism of Cyclic Esters Polymerization Initiated with Tin(II) Octoate. Polymerization of ε-Caprolactone and L, L-Lactide Co-initiated with Primary Amines. *Macromolecules* **2005**, *38*, 8170–8176. [CrossRef]

22. Trollsås, M.; Hedrick, J.L.; Mecerreyes, D.; Dubois, P.; Jérôme, R.; Ihre, H.; Hult, A. Highly Functional Branched and Dendri-Graft Aliphatic Polyesters through Ring Opening Polymerization. *Macromolecules* **1998**, *31*, 2756–2763. [CrossRef]

23. Trathnigg, B.; Yan, X. Copolymer Analysis by SEC with Dual Detection. Coupling of Density Detection with UV and RI Detection. *Chrornatographia* **1992**, *33*, 467–468. [CrossRef]

24. Zhao, Y.L.; Cai, Q.; Jiang, J.; Shuai, X.T.; Bei, J.Z.; Chen, C.F.; Xi, F. Synthesis and thermal properties of novel star-shaped poly (L-lactide) s with starburst PAMAM–OH dendrimer macroinitiator. *Polymer* **2002**, *43*, 5819–5825. [CrossRef]

25. Shaver, M.P.; Cameron, D.J.A. Tacticity control in the synthesis of poly (lactic acid) polymer stars with dipentaerythritol cores. *Biomacromolecules* **2010**, *11*, 3673–3679. [CrossRef] [PubMed]

26. Palacio, J.; Orozco, V.H.; López, B.L. Effect of the molecular weight on the physicochemical properties of poly (lactic acid) nanoparticles and on the amount of ovalbumin adsorption. *J. Braz. Chem. Soc.* **2011**, *22*, 2304–2311. [CrossRef]

27. Michell, R.M.; Müller, A.J.; Boschetti-de-Fierro, A.; Fierro, D.; Lison, V.; Raquez, J.-M.; Dubois, P. Novel poly (ester-urethane)s based on polylactide: From reactive extrusion to crystallization and thermal properties. *Polymer* **2012**, *53*, 5657–5665. [CrossRef]

28. Kucharczyk, P.; Pavelkova, A.; Stloukal, P.; Sedlarík, V. Degradation behaviour of PLA-based polyesterurethanes under abiotic and biotic environments. *Polym. Degra. Stab.* **2016**, *129*, 222–230. [CrossRef]

29. Schäfer, A.; Seibold, S.; Lohstroh, W.; Walter, O.; Döring, M.J. Synthesis and properties of flame-retardant epoxy resins based on DOPO and one of its analog DPPO. *Appl. Polym. Sci.* **2007**, *105*, 685–696. [CrossRef]

30. Hao, X.; Kaschta, J.; Hu, X.; Pan, L.Y.; Schubert, D.W. Entanglement network formed in miscible PLA/PMMA blends and its role in rheological and thermo-mechanical properties of the blends. *Polymer* **2015**, *80*, 38. [CrossRef]

Article

Exploring the Contribution of Two Phosphorus-Based Groups to Polymer Flammability via Pyrolysis–Combustion Flow Calorimetry

Rodolphe Sonnier [1,*], Belkacem Otazaghine [1], Christelle Vagner [2,3,4], Frédéric Bier [5,6], Jean-Luc Six [5,6], Alain Durand [5,6] and Henri Vahabi [2,3]

[1] Centre des Matériaux des Mines d'Alès (C2MA), IMT Mines Alès-6, Avenue de Clavières, CEDEX 30319 Alès, France; belkacem.otazaghine@mines-ales.fr
[2] Laboratoire Matériaux Optiques, Photonique et Systèmes (LMOPS), CentraleSupélec, Université de Lorraine, F-57000 Metz, France; christelle.vagner@univ-lorraine.fr (C.V.); henri.vahabi@univ-lorraine.fr (H.V.)
[3] Laboratoire Matériaux Optiques, Photoniques et Systèmes, CentraleSupélec, Université Paris-Saclay, 57070 Metz, France
[4] Aix Marseille Univ, French National Center for Scientific Research (CNRS), MADIREL UMR 7246, F-13397 Marseille, France
[5] Laboratoire de Chimie Physique Macromoléculaire (LCPM), Université de Lorraine, UMR7375, F-54001 Nancy, France; frederic.bier@gmail.com (F.B.); jean-luc.six@univ-lorraine.fr (J.-L.S.); alain.durand@univ-lorraine.fr (A.D.)
[6] Laboratoire de Chimie Physique Macromoléculaire (LCPM), French National Center for Scientific Research (CNRS), UMR7375, F-54001 Nancy, France
* Correspondence: rodolphe.sonnier@mines-ales.fr

Received: 30 August 2019; Accepted: 10 September 2019; Published: 12 September 2019

Abstract: From a set of around 100 phosphorus-containing polymers tested in pyrolysis–combustion flow calorimetry, the contributions to flammability of two phosphorus-containing pendant groups (called 9,10-dihydro-9-oxa-10-phosphaphenanthrene-10-oxide (DOPO) and PO_3) were calculated using an advanced method previously proposed and validated. The flammability properties include total heat release (THR) and heat release capacity (HRC) measured in standard conditions, i.e., anaerobic pyrolysis and complete combustion. The calculated contributions are in good agreement with the main modes of action of both phosphorus groups, i.e., flame inhibition for DOPO and char promotion for PO_3. Moreover, the results provide first conclusions about the cooperative interaction between phosphorus and nitrogen, as well as the influence of the architecture of tested co-polymers.

Keywords: polymer flammability; van Krevelen approach; group contributions; pyrolysis–combustion flow calorimetry; phosphorus-containing flame retardant

---

1. Introduction

Phosphorus is currently considered as a key element to develop flame-retardant (FR) materials. Indeed, it can act as a char promoter in the condensed phase and/or as a flame inhibitor in the gaseous phase [1]. Numerous phosphorus compounds are currently available, mainly as additives for polymers. Nevertheless, phosphorus groups can also be chemically incorporated into thermoplastic chains or thermoset networks [2,3]. This so-called reactive approach is expected to enhance the durability of materials by preventing the migration of FR outside the polymer matrix [4]. Moreover, the transparency, as well as the mechanical properties, may be more easily maintained through the reactive approach, since phosphorus additives generally behave as plasticizers and lower the glass transition temperature of final materials compared to that of bulk polymers. Even if the phosphorus-containing groups chemically bonded to polymers can also affect the glass transition temperature, this effect may be significantly reduced in comparison with phosphorus additives. Finally, several studies suggested that the reactive approach was more efficient than the additive one [5–9]. A review paper about the flame retardancy of phosphorus-containing polymers was recently published [10].

The efficiency of phosphorus, as well as its mode of action in the condensed or gaseous phase, depends on a couple of parameters, including its oxidation state [11–14] and the host polymer [15]. When phosphorus is covalently bonded to the polymer, its exact position on the main chain also has an effect. It can be located in the backbone as in polyphosphazenes, which exhibit high flame retardancy [16]. Nevertheless, most often, phosphorus groups are positioned as pendant groups. Thus, selecting the exact chemical structure of a phosphorus-containing monomer may be an efficient way to increase the FR efficiency. Additionally "synergism" with other chemical groups (aromatic rings, as well as "flame-retardant elements" such as nitrogen, sulfur, etc.) is also postulated, and many attempts were made to prepare flame-retardant polymers combining phosphorus and nitrogen [17,18], sulfur [19,20], silicon [21,22], and bromine [23], as well as calcium or zinc [24–26]. Nevertheless, "synergism" is most often difficult to confirm.

Among the phosphorus-containing groups incorporated into polymers, 9,10-dihydro-9-oxa-10-phosphaphenanthrene-10-oxide (DOPO), as well as its derivatives [8,27–31], and phosphonate groups [9,17,32–37] are probably the most widely studied. DOPO mainly acts as a flame inhibitor, while phosphonate groups promote charring with no or little effect in the gaseous phase. In most cases, these groups were added as pendant groups even if some attempts were carried out to incorporate phosphonate groups within the polymer backbone [32,33].

Predicting the flammability of polymers was recently successful on the basis of molecular structure of repeat units using experimental data obtained by pyrolysis–combustion flow calorimetry (PCFC) and considering a simplistic model based on the additivity of molar groups contributions [38,39]. This approach was originally proposed by van Krevelen [40] and was firstly applied to PCFC data by Lyon et al. [41,42]. A database gathering the contributions for about 45 chemical groups was proposed and validated for about 140 thermoplastics and thermosets [38,39]. Such a model should help chemists to consider polymeric structures to be synthetized in order to obtain low flammability without proceeding through a long and expensive trial-and-error process. Moreover, it should help to identify interesting structures. Indeed, interactions between chemical groups may lead to a discrepancy between the experimental and calculated flammability properties of specific polymers, while the model is additive and does not consider contributions from interactions.

Nevertheless, no satisfactory contribution was proposed for phosphorus-containing groups, mainly because too few phosphorus-containing polymers were tested [38,43]. Moreover, the phosphorus content in such polymers is usually low and, therefore, it is difficult to accurately calculate the contributions of phosphorus-containing groups to flammability.

In this work, the contributions of two phosphorus-containing pendant groups (namely, DOPO and PO_3 groups) were calculated from a large set of around 100 polymers. These groups were chosen because they are widely used to impart flame retardancy to new polymers. Furthermore, they act according to different modes of action. Moreover, the content of both phosphorus-containing groups reaches high values in many polymers. The calculated contributions of both groups allow for reasonable fitting of the flammability properties of the polymers considered at the molecular scale. This approach also allows the formulation of preliminary conclusions about the interaction of two or more flame-retarding elements simultaneously present and about the architecture of co-polymers. To the best of the authors' knowledge, this is the first time that the contributions of phosphorus-containing groups are satisfactorily calculated from such a large set of polymers.

2. Materials and Methods

Around 100 phosphorus-containing polymers grouped into eight series (from A to H) were studied (Table 2). For each series, the differences between the polymers were as follows: the nature of one or several co-monomers and/or the ratio between the co-monomers. Details about most of these polymers can be found elsewhere (references are given in the last column of Table 2). Their flammability was already reported in various articles except for polymers from series A and C.

The detailed synthesis procedure of polymers from series A is described in detail elsewhere [44]. The two-step synthesis strategy used the Atherton–Todd reaction [45,46]. The first step consisted of, firstly, a radical reaction which activated an alkenol (of variable length between 3 and 11 carbon atoms) in the presence of a radical initiator (azobisisobutyronitrile, AIBN) at 70 °C. The second step was the introduction of the methacrylic function and was realized by the nucleophilic substitution between the previous DOPO-alcohol and methacryloyl chloride. The DOPO-alkan-methacrylates were named DOPO-MnP, where n is the number of methylene groups contained in the aliphatic spacer.

Table 1. List of groups studied in the present work and their estimated contributions to flammability. HRC—heat release capacity.

Group	Number of Polymers	Molar Mass (g/mol)	Contribution to			
			THR (kJ/g)	HRC (J/g·K)	Δh (kJ/g)	Char (g/g)
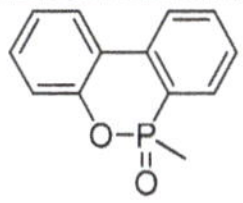 (DOPO structure)	35	215	27	270	27.8	0.02
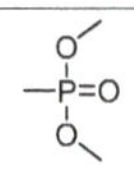 (phosphonate —P=O structure)	57	79	-8	−400	-0.7	0.20
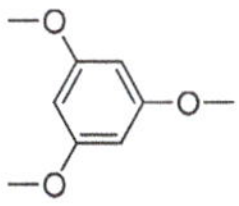 (trioxybenzene structure)	29	123	0	−350	23	0.62
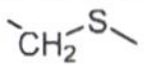 (—CH$_2$—S— structure)	12	46	5	−300	22.8	0.48

Table 2. List of polymers tested in the present study. DOPO—9,10-dihydro-9-oxa-10-phosphaphenanthrene-10-oxide; MAPC1—dimethyl(methacryloxy)methyl phosphonate; MANP2C3—2-(bis((dimethoxyphosphoyl)methyl)amino)ethyl methacrylate.

Series	Description	Number of Tested Polymers	Structure and Content (wt.%) of the Phosphorus Group	Reference
A	DOPO-containing acrylate and methacrylate co-polymers	35	DOPO Up to 63 wt.%	This work
B	MAPC1 and MANP2C3-containing methacrylate random co-polymers	6	PO$_3$ Up to 41 wt.%	[35]
C	MAPC1-containing methacrylate block co-polymers	7	PO$_3$ Up to 38 wt.%	[47]
D	Phosphonate-containing epoxy thermosets (including trioxybenzene group **)	17	PO$_3$ Up to 14 wt.%	[36]
E	Phosphonate and sulfur-containing epoxy thermosets (including trioxybenzene and methylene sulfide groups **)	12	PO$_3$ Up to 8 wt.%	[9]
F	Phosphonate-containing co-polymers	9	PO$_3$ Up to 60 wt.%	[38]
G	Polymers from phosphorus-modified styrene monomers	7	PO$_3$ Up to 35 wt.%	[17,37]
H *	Polymers from phosphorus-modified styrene monomers	4	NHPO and NHPO$_3$ Up to 39 wt.%	[17]

* Only total heat release (THR) values were considered in the present article; ** See Table 1.

Co-polymerization of the DOPO-containing monomer and methyl methacrylate was carried out in a test tube equipped with a three-way and a magnetic stirrer. The co-monomers and the initiator (AIBN, 1 wt.% compared to monomer) were solubilized in dimethyl sulfoxide (DMSO). Reaction time was fixed at 15 h and the temperature was 80 °C. Co-polymers were recovered by precipitation into diethyl ether. Finally, samples were dried 24 h in an oven at reduced pressure at 40 °C until constant weight.

The structures of various phosphorus-functionalized monomers listed in Table 2 (series A, B, and C) are provided in Figure 1. Polymers from series A were statistical co-polymers of DOPO-containing monomers and methyl methacrylate (MMA) prepared using radical polymerization. DOPO-containing monomers were of acrylic (in one case) and mainly methacrylic type with DOPO as a pendant group. The final content of the phosphorus-containing monomer was up to 50% in moles. Homopolymers of DOPO-containing monomers were also prepared following the same procedure. Note also that the oxidation state of the phosphorus is not the same for all the co-polymers from series A (compare DOPO–2-(6-oxidodibenzo[c,e][1,2]oxaphosphinin-6-yl)oxy)ethyl methacrylate (HEMA) and others). Nevertheless, as shown below, the contributions calculated for this group allow predicting the flammability properties of all the co-polymers from series A, regardless of the true oxidation state of phosphorus.

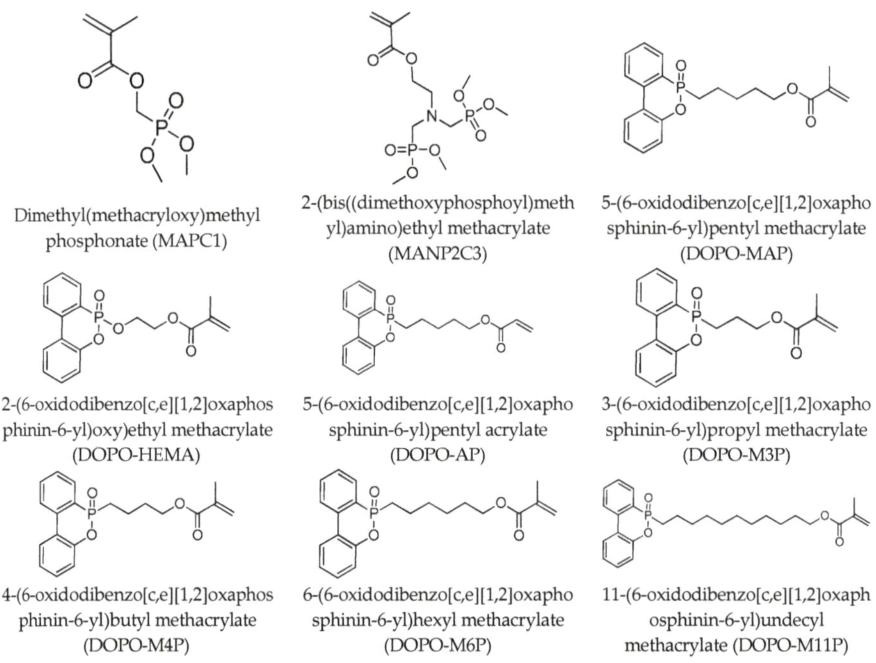

Figure 1. Structure of 9,10-dihydro-9-oxa-10-phosphaphenanthrene-10-oxide (DOPO)-functionalized monomers, MAPC1 and MANP2C3.

Flammability of the polymers listed in Table 2 was analyzed using PCFC (from FTT, United Kingdom) under standard conditions, i.e., anaerobic pyrolysis from 25 to 750 °C at 1 °C/s in nitrogen and complete combustion in an excess of oxygen at 900 °C [48]. O_2 and N_2 volume fractions in the combustor were fixed at 0.2 and 0.8, respectively. The sample weight was typically 2–3 mg so that oxygen was never fully consumed. Moreover, the sample was considered as thermally thin.

The total heat release (THR) corresponds to the area under the heat release rate curve. The heat release capacity (HRC) generally corresponds to the peak of heat release rate (pHRR) divided by the heating rate. However, in some cases, several peaks can be observed. In such a case, the sum of the

HRR peaks after deconvolution carried out using the FTT software is considered (sumHRC). When the different peaks do not overlap, the deconvolution is easy and unambiguous. For the cases where several peaks overlap, sumHRC was determined as previously described by summing the minimum number of Gaussian, Lorentzian, asymmetric Gaussian or Lorentzian, or asymmetric Gaussian–Lorentzian hybrid peaks needed to fit the HRR curve with an accuracy of at least 95% [41]. Obviously, the choice of the number of peaks influences the sumHRC.

Equations (1) and (2) explain how the total heat release (THR) and heat release capacity (HRC) of a given polymer can be calculated on the basis of the structure of its repeat unit.

$$\text{THR} = \sum_i w_i \times \text{THR}_i, \tag{1}$$

$$\text{HRC} = \sum_i w_i \times \text{HRC}_i, \tag{2}$$

where THR_i and HRC_i reflect the contributions of group i to THR and HRC, respectively, and w_i is the weight fraction of group i in the polymer.

Most of the polymers studied in this article have a low or negligible char yield (series A, for example). In some cases, these data were not recorded when PCFC analyses were carried out (series B, C, and F). Therefore, the char yield was not calculated and compared to experimental data. Nevertheless, the contribution to char (μ) of the new groups considered here can be calculated using Equation (3), from the comparison between the contribution to THR and the heat of complete combustion (Δh), as described in previous articles [38,39]. Walters et al. showed that the char composition is close to C_5H_2 in most cases [49]. The energy released by the complete pyrolysis and combustion of such char Δh_{char} is then 37.2 kJ/g. When the contribution to THR is significantly different from the heat of complete combustion, it means that contribution to char is significant. Δh is calculated using Huggett's relation [50] considering the complete pyrolysis and combustion of the whole polymer structure.

Δh is calculated without considering the oxidation of nitrogen atoms since oxidation of nitrogen is believed to occur at much higher temperature [51].

$$\mu = \frac{\Delta h - \text{THR}}{\Delta h_{char}}. \tag{3}$$

This method is simplistic because it considers that all the chars have the same composition. Nevertheless, the method allows for correct prediction of the THR and the char content of the polymers from only one parameter: the contribution to THR. This was the case in a previous work for a series of around 30 thermosets including high-charring polymers [39].

The calculation of the heat of complete combustion for phosphorus-containing groups requires information about the oxidation state of phosphorus after oxidation. Phosphorus was considered to be fully oxidized into P_2O_5, i.e., 2.5 atoms of oxygen for one atom of phosphorus. Lyon et al. proposed that phosphorus species were converted into H_3PO_4, i.e., $0.5P_2O_5 + 1.5H_2O$, which was similar to the calculation used in this work [52]. This leads to a slightly negative heat of complete combustion in the case of the PO_3 group since this group is made up of three atoms of oxygen for one atom of phosphorus.

More details about the method to build the database of calculated group contributions step-by-step can be found in previous articles [38,39].

3. Results and Discussion

Some preliminary precisions about the model and its limitations are needed. Firstly, the model allows calculating negative contributions. From a physical point of view, negative contributions are meaningless. Indeed, one polymer containing only one group exhibiting negative contributions would present itself negative flammability properties. Nevertheless, if the content of the group having negative contributions is not too high, the flammability properties of the polymer remain positive.

In that case, such a group may be considered as acting to its neighboring groups. It modifies their degradation pathway and reduces their contribution to flammability. When such an interaction is more or less systematic whichever the neighbors, it can be included into the own contributions of the group.

Of course, deviations between experimental and calculated values are unavoidable. A first reason is that the calculation of contributions evolves and depends on the set of studied polymers. However, the main reason is that the model is very simplistic. The decomposition pathway is never exactly as a sum of independent steps corresponding each one to a chemical group (which is the meaning of a model based on additive contributions). When the deviations are limited, it means that such a model remains a reasonable assumption. When the deviations are high, it can be assumed that interactions occur between some groups present in the polymer. In order to help to identify them, such interactions may be presented as another interest of the model.

The contributions of new groups calculated in this study are listed in Table 1. It should be noted that the contributions of trioxybenzene group (group 3 in Table 1) were already calculated in a previous work [39]. The contribution to THR found in the present work is the same but the contribution to HRC differs significantly (previous value was 100 J/g·K and, in the present work, the calculations led to −350 J/g·K). This discrepancy can easily be justified when considering that, in the previous article, the contribution was calculated using a series of only three polymers as compared to 29 in the present work. Thus, the previously proposed value of that contribution must be considered carefully. The new contribution appears more reliable because it allows for satisfactory prediction of the HRC of 29 polymers (series D and E), including the three phosphorus-free polymers used in the previous work.

Tentative contributions of the PO$_3$ pendant group were also calculated in a previous work [38] but these values were reported as unsatisfactory. The new contributions shown in Table 1 allow for a more accurate prediction of the flammability properties of 57 polymers (including most of the polymers already studied).

Similarly, the contributions of methylene sulfide group (group 4 in Table 1, from series E) must be considered as tentative. Indeed, even if this group is present in 12 polymers, its weight fraction is reduced in all cases and, thus, its contribution has low influence on the predicted values. Therefore, we did not discuss these values in detail.

This illustrates that the database is still under construction, and some contributions can be modified in the future depending on the availability of new experimental data.

The contributions of DOPO were calculated from a large range of 35 homo- and co-polymers (series A). Its weight fraction ranged from 0.27 to 0.63. Therefore, its influence on the heat release may be very significant, which should be in favor of the accuracy of the estimated contributions.

Nevertheless, the THR and HRC of all these 35 polymers were in the same range (Figures 2 and 3). HRC ranged from 300 to 500 J/g·K and THR ranged from 20 to 30 kJ/g. This means that the contributions of DOPO are similar to the contributions of the other groups present in the polymers. Indeed, the contributions to HRC and THR are respectively 270 J/g·K and 27 kJ/g, which are relatively high values. Therefore, DOPO cannot be considered as an efficient flame-retardant group considering the fact that values were measured in PCFC in standard conditions.

DOPO is well known to usually be a poor char promoter but efficient as a flame inhibitor. However, in standard conditions, combustion is complete. Indeed, the temperature of the combustor is 900 °C, while the combustion needs a temperature lower than 700 °C to be incomplete for most polymers. To the best of the authors' knowledge, only one polymer, poly(4-bromostyrene), exhibited incomplete combustion at 900 °C [53,54]. Therefore, the efficiency of flame inhibitors as DOPO is underestimated when using these conditions, as already proven in a previous work [53]. As an example, 1 wt.% phosphorus provided by the incorporation of a DOPO-containing group significantly improved the flame retardancy of polyamide 11 according to limiting oxygen index (LOI) and UL94 tests even if flammability at microscale was not modified [31].

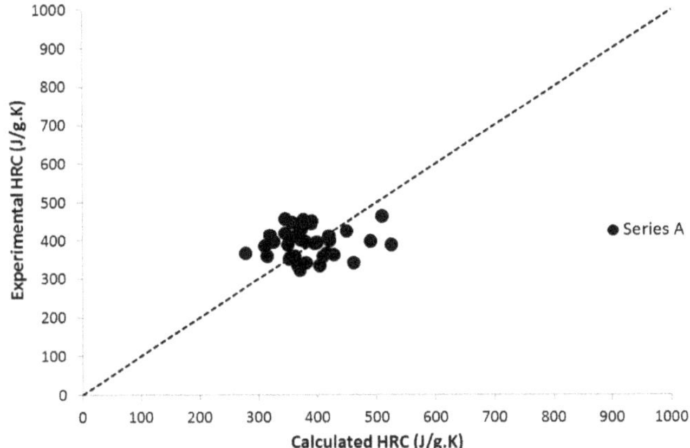

Figure 2. Experimental versus calculated heat release capacity (HRC) for 35 DOPO-functionalized polymers.

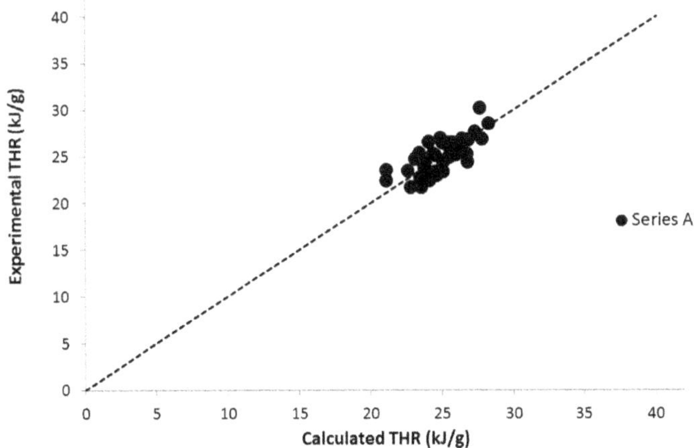

Figure 3. Experimental versus calculated total heat release (THR) for 35 DOPO-functionalized polymers.

The calculated THR and HRC values for phosphonated polymers (series B, C, D, E, and F) are shown in Figures 4 and 5. The weight fraction of the PO_3 group changes greatly from one polymer to another, but it reaches up to 0.4 for homopolymer poly(MANP2C3). The agreement between calculated and experimental values can be considered as quite satisfactory. Some deviations from the dotted line are discussed below and may be attributed to the influence of the co-polymer architecture. In previous work, we also added three polymers for which the experimental HRC values were graphically obtained from the literature. These polymers were not considered in the present work.

It can be noted that the PO_3 pendant group is much better than DOPO at improving flame retardancy. The PO_3 pendant group has a significant effect on HRC, i.e., it reduces heat release rate to a great extent. Indeed, phosphonate is well known as a char promoter, modifying the decomposition mechanisms in the condensed phase (i.e., during the pyrolysis step). However, the contribution to THR is negative mainly because the PO_3 group does not contain carbon atoms and its heat of combustion is low. Its contribution to char remains relatively limited (0.2 g/g). It is expected that this contribution would be much higher if this group was incorporated into the polymer backbone.

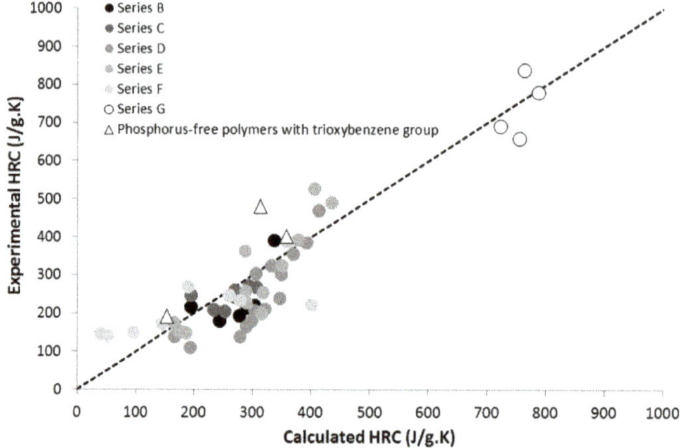

Figure 4. Experimental versus calculated HRC for polymers bearing PO$_3$ pendant groups.

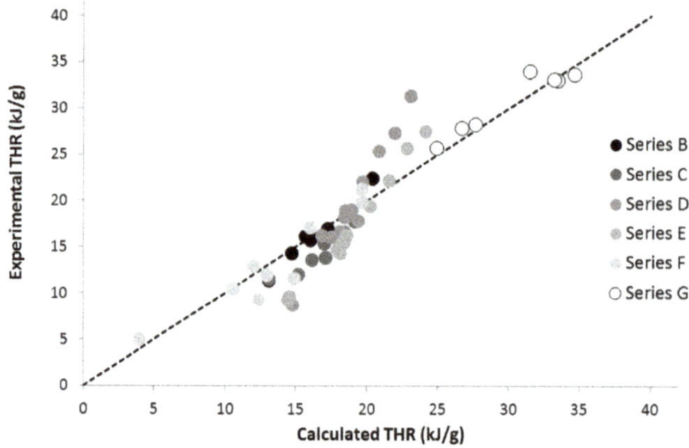

Figure 5. Experimental versus calculated THR for polymers bearing phosphonate pendant groups.

Figure 6 shows the contributions to HRC versus the contributions to THR for all groups already studied, i.e., 47 groups. The complete list of these groups and their corresponding contributions are available in previous papers [38,39]. Both DOPO and PO$_3$ pendant groups follow the same rough tendency between these contributions. The higher the contribution to THR is, the higher the contribution to HRC is. This is not unexpected, but this rough tendency allows for identification of some exceptions, such as >C< and –CH< (high contribution to THR and low contribution to HRC) or >CH–O– (low contribution to THR and high contribution to HRC).

Figure 7 shows the contribution to THR versus the heat of combustion calculated using Huggett's relation. When both values are close, it means that the group does not significantly contribute to char. This is the case of DOPO. The PO$_3$ pendant group contributes only moderately to char as already discussed, especially when comparing it with heteroaromatic groups (see some examples in Figure 7). Therefore, this data point is not far away from the dotted line.

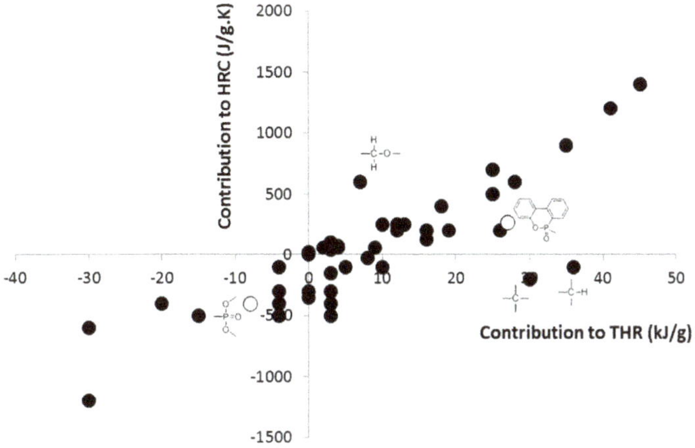

Figure 6. Contribution to HRC versus contribution to THR for 47 groups.

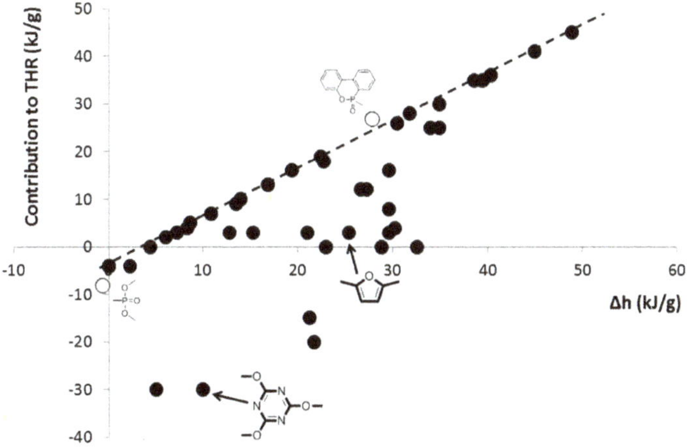

Figure 7. Contribution to THR versus Δh calculated from the Huggett relation for 47 groups.

3.1. Cooperative Interactions

As already mentioned, "synergism" between "flame-retardant" elements such as phosphorus, nitrogen, or sulfur is often claimed, but the evidence for cooperative interactions is rather scarce [55]. Moreover, it is unclear whether such interactions should occur only when both elements are directly linked through a covalent bond or even when both elements are not directly linked. In this work, several polymers containing N and P or S and P elements were studied but without direct bonding between these atoms. At least one carbon atom was present between N and P or S and P atoms in all corresponding polymers. In the case of sulfur-containing polymers, the contribution of CH_2–S was not calculated independently (i.e., in phosphorus-free polymers); therefore, it is not possible to conclude anything.

The contributions of both $-CH_2-N<$ and PO_3 pendant groups were calculated independently. It appears that these contributions also allow correctly predicting the flammability properties of the N- and P-containing polymers (containing MANP2C3 monomers, from series B) (Figure 8). Note that the weight fractions of the N- and P-containing groups are significant, especially for the homopolymer poly(MANP2C3). Therefore, it is assumed that no cooperative interaction occurs in these polymers when N and P atoms are not directly linked. Vahabi et al. also calculated the contributions of the PO_3 group in MAPC1- and MANC2P3-containing co-polymers using another approach [35]. They also concluded that the contributions to effective heat of combustion (EHC) and char yield were similar in both co-polymers (i.e., the contribution to THR was also similar). Nevertheless, the contributions to HRC were −258 J/g·K and −549 J/g·K for PO_3 groups in MAPC1-containing co-polymers and in MANC2P3-containing co-polymers, respectively. However, the fitting of calculated values with experimental ones was less satisfying, especially for MAPC1-containing co-polymers. It is noteworthy that these values are quite close to the contribution calculated in the present work (−400 J/g·K).

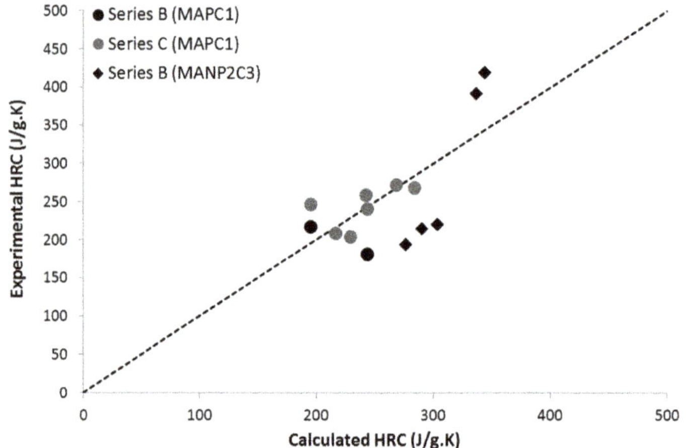

Figure 8. Experimental versus calculated HRC for polymers containing PO_3 pendant groups from series B and C (i.e., methacrylate co-polymers including MAPC1 or MANC2P3 monomers).

This first example illustrates one main advantage of the present database based on the Van Krevelen method. Indeed, a cooperative effect can be highlighted or rejected on the basis of a quantitative assessment.

Previously, Dumitrascu and Howell synthetized and analyzed other polymers containing phosphorus groups including covalent bonds between N and P [17]. Nevertheless, the number of polymers was too limited. Moreover, the article provides only THR but not sumHRC values. Therefore, these polymers were not plotted in the previous figures (series H). Nevertheless, we calculated the contribution of these groups to THR to properly fit the experimental THR. The correlations between experimental and calculated THR, as well as the corresponding contributions to THR, are shown in Figure 9. Data for other polymers containing N and P or S and P atoms (without direct bonding—from series B and E) were also added.

To correctly fit the experimental THR, the contributions of $NHPO_3$ and NHPO would be close to −17 and −10 kJ/g, respectively. These values are very low and correspond to contributions to char close to 0.45. Indeed, the residue contents for these polymers are significantly higher than for their counterparts containing only phosphorus groups. Based on these preliminary results, it seems that a high flame-retardant effect may be expected from groups containing an N–P bond.

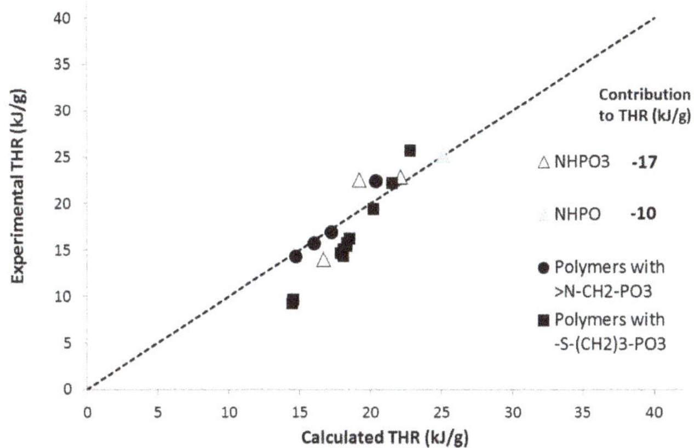

Figure 9. Experimental versus calculated THR for various polymers (from series B, E, and H).

3.2. Influence of the Detailed Structure of Co-Polymers

Co-polymers of MMA and MAPC1 were synthetized by Vahabi et al. and Canniccioni et al. [35,47]. The former prepared random co-polymers while the latter synthesized block co-polymers. Moreover, we also studied physical blends of poly(MMA) (PMMA) and poly(MAPC1) (PMAPC1). The THR of all these materials can be predicted accurately using the contributions to THR previously calculated. In other words, there is no difference in THR values between random and block co-polymers or physical blends if the weight fractions of MMA and MAPC1 are the same.

Prediction of HRC is also rather reasonable but not perfect for block and random co-polymers (Figure 4). Figure 10 plots the experimental HRC versus MAPC1 content in the co-polymers as in physical blends in order to highlight possible differences between these materials. A slightly negative deviation is highlighted for co-polymers, especially the random co-polymer. For a similar composition, HRC is the highest for physical blend, and the lowest for co-polymers. Nevertheless, this deviation appears quite limited and cannot be considered as significant on the basis of experimental uncertainties. Further investigations are needed to draw clear conclusions on that point.

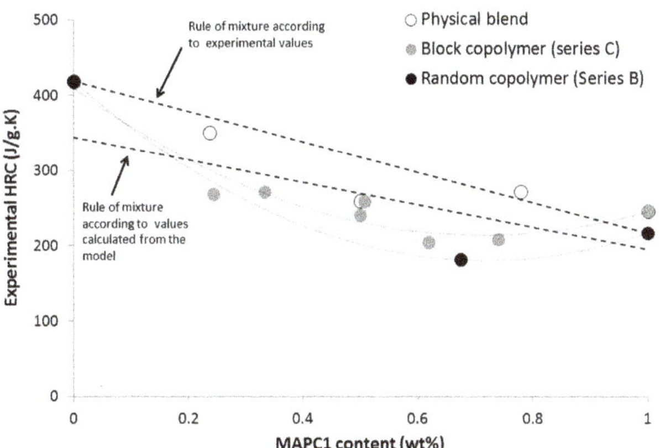

Figure 10. Experimental HRC versus MAPC1 content for various polymers (from series B and C).

Interestingly, the lowest value for series B and C (i.e., random and block co-polymers) was not obtained for the homopolymer PMAPC1 but for an MAPC1 weight fraction close to 0.6–0.7.

This fraction corresponds to a ratio between the acrylate COO group and PO$_3$ group close to 2. This ratio is exactly the same as the one obtained in a previous work on phosphorus-containing oligomers and polymers [43]. Indeed, in this previous work, a method was presented to assess the interactions between groups based on PCFC results from another set of molecules and macromolecules. Cooperative interaction was the highest for this ratio between ester and PO$_3$ groups.

3.3. Comparison of the Contributions Calculated from Other Works

It is interesting to compare the calculated contributions in the present work to those already proposed (Table 3). The comparison must be considered carefully because different approaches were followed according to the references. Moreover, some contributions were calculated with a very limited number of polymers (for example, in this work, NHPO and NHPO$_3$). It is noteworthy that the contributions to THR and HRC of phosphorus groups without carbon atoms can be slightly positive or negative, confirming the flame-retardant effect of phosphorus (Figures 11 and 12). The only exception is the contribution to HRC of the PO$_3$ group calculated from a preliminary work [43]. This may be due to the small size of some molecules. The decomposition pathway of such molecules may be different from that of polymers.

Table 3. List of phosphorus-containing groups and the corresponding contributions to flammability.

Group	Contribution to THR (kJ/g)	Contribution to HRC (J/g·K)	Reference
(phosphate group, —P=O with O substituents)	−8	−400	This work
	−5	−100	[38]
	−0.2 *	−258	[35]
	1.4 *	−549	[35]
	0.5	500	[43]
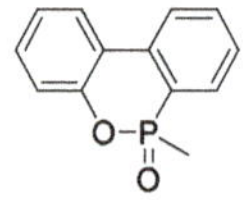 (DOPO group, O–P=O in biphenyl structure)	27	270	This work
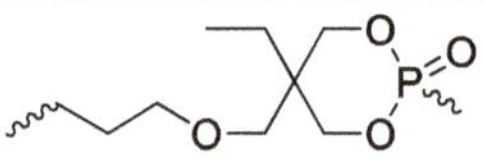 (cyclic phosphate, O–P=O bridged structure)	20	400	[43]
(phosphate, O–P(=O)(O)–O)	/	−807	[42]
(phosphazene, —N=P— with O substituents)	/	−179	[42]
(triaryl phosphine oxide, aryl–P(=O)–aryl)	/	279	[42]
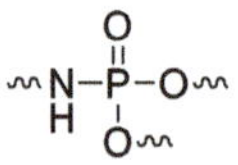 (N–P(=O)–O phosphoramidate)	−17	/	This work
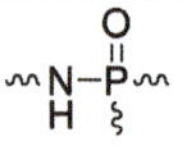 (N–H–P=O phosphonamide)	−10	/	This work

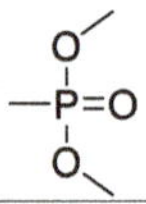

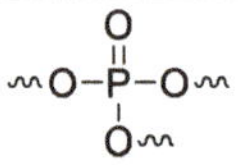

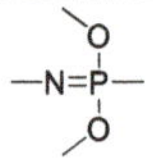

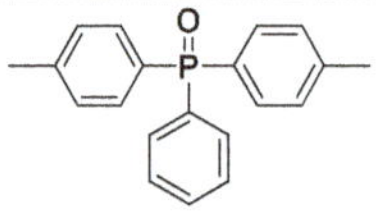

* Calculated from contributions to effective heat of combustion (EHC) and char.

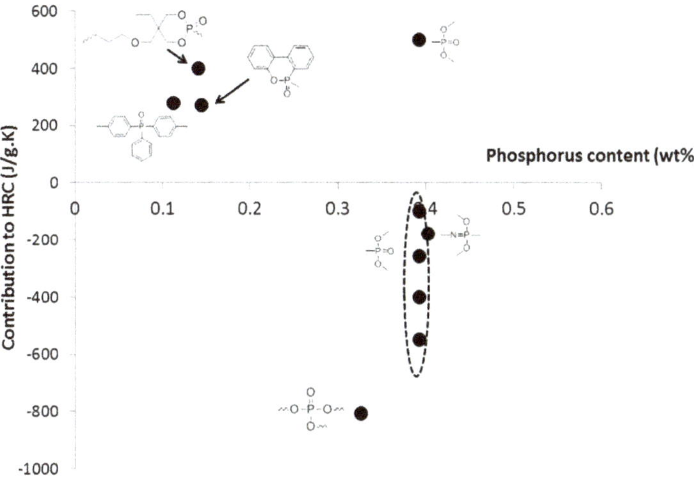

Figure 11. Contributions to HRC for various phosphorus-containing groups according to different works.

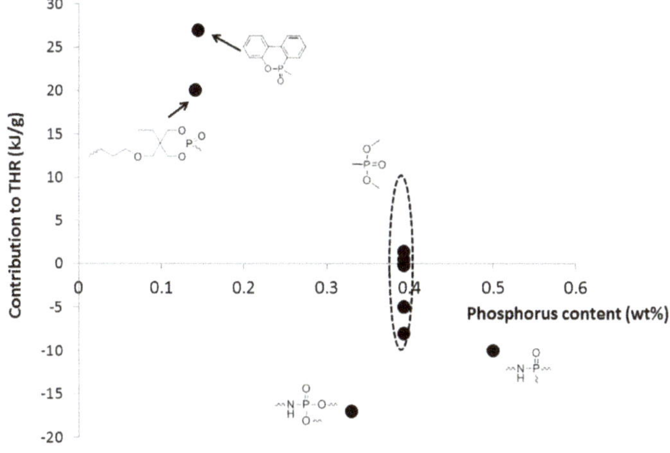

Figure 12. Contributions to THR for various phosphorus-containing groups according to different works.

When the phosphorus group contains a significant carbon content, its contributions are notably higher (Figures 11 and 12). The effect of phosphorus is then "diluted" and is hardly worthy of being highlighted. Nevertheless, it is noteworthy that a non-aromatic heterocycle can be decomposed into the smallest groups in the present model. Therefore, the contributions of the dioxaphosphorinane group studied in a previous work [43] can also be calculated from the contributions of small groups including the PO_3 group (see Figure 13 for the structure fragmentation of a dioxaphosphorinane group). When using these contributions calculated from the present work (given in Table 1 for the PO_3 group) and previous works (shown in Reference [38] for other groups), the contributions to THR and HRC of dioxaphosphorinane are 18.9 kJ/g and 433 J/g·K, respectively. This is in very good agreement with the contributions calculated previously: 20 kJ/g and 400 J/g·K considering the dioxaphosphorinane as a whole.

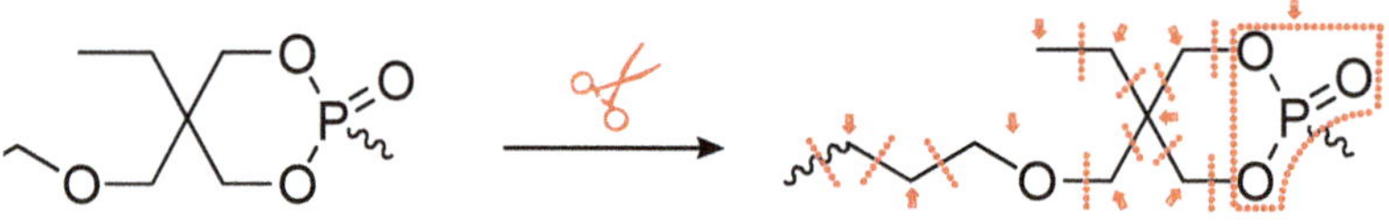

Figure 13. Structure fragmentation of dioxaphosphorinane group into smaller groups.

4. Conclusions

This attempt to calculate the contributions to flammability of phosphorus groups is the first considering nearly 100 polymers. Taking into account the increasing interest of phosphorus-containing FR compounds, such correlations become highly suitable. The calculated contributions are in good agreement with the main mode of action of DOPO and PO_3 groups. DOPO is mainly a flame inhibitor and its contributions are high when combustion is complete as in standard PCFC conditions. On the contrary, the PO_3 pendant group is a char promoter, modifying the decomposition rate. Its contribution to HRC is very low, evidencing its flame-retardant effect. Its contribution to char remains limited.

The model allows calculating the flammability properties according to the additivity of molar contributions without considering any interactions between groups. Therefore, the deviation between predicted and experimental values highlights a possible cooperative or antagonistic effect. First!conclusions can be drawn on the basis of the present results. The combination of N and P atoms does not act cooperatively when these atoms are not directly bonded. The architecture of co-polymers (random versus block co-polymers) has no or limited effect on flammability. Interactions between ester and PO_3 groups may be beneficial to reduce flammability. Even if these conclusions must be considered with caution and require further investigations, these examples illustrate the usefulness of the model and the related database.

Author Contributions: Conceptualization, R.S., B.O. and H.V.; methodology, R.S. and B.O.; validation, R.S., F.B., J.-L.S. and A.D.; formal analysis, R.S.; investigation, R.S., F.B., J.-L.S., A.D.; resources, R.S., F.B., J.-L.S., A.D.; writing—original draft preparation, R.S.; writing—review and editing, R.S., B.O., H.V. and C.V.; visualization, R.S., B.O. and H.V.; supervision, R.S.; project administration, R.S. and H.V.

Funding: This research received no external funding.

Acknowledgments: The authors thank Loïc Dumazert for his help in carrying out analyses and Claire Negrell, Ghislain David, Raphaël Ménard, and coworkers for the synthesis of the many polymers studied.

Conflicts of Interest: The authors declare no conflict of interest.

References

1. Schartel, B. Phosphorus-based Flame Retardancy Mechanisms—Old Hat or a Starting Point for Future Development? *Materials* **2010**, *3*, 4710–4745. [CrossRef] [PubMed]
2. Illy, N.; Fache, M.; Menard, R.; Negrell, C.; Caillol, S.; David, G. Phosphorylation of bio-based compounds: The state of the art. *Polym. Chem.* **2015**, *35*, 6257–6291. [CrossRef]
3. Bauer, K.; Tee, H.; Velencoso, M.; Wurm, F. Main-chain poly(phosphoester)s: History, syntheses, degradation, bio-and flame-retardant applications. *Prog. Polym. Sci.* **2017**, *73*, 61–122. [CrossRef]
4. Vahabi, H.; Sonnier, R.; Ferry, L. Effects of ageing on the fire behaviour of flame-retarded polymers: A review. *Polym. Int.* **2015**, *64*, 313–328. [CrossRef]
5. Price, D.; Pyrah, K.; Hull, T.R.; Milnes, G.J.; Ebdon, J.R.; Hunt, B.J.; Joseph, P.; Konkel, C.S. Flame retarding poly(methyl methacrylate) with phosphorus-containing compounds: Comparison of an additive with a reactive approach. *Polym. Degrad. Stab.* **2001**, *74*, 441–447. [CrossRef]
6. Price, D.; Pyrah, K.; Hull, T.R.; Milnes, G.J.; Ebdon, J.R.; Hunt, B.J.; Joseph, P. Flame retardance of poly(methyl methacrylate) modified with phosphorus-containing compounds. *Polym. Degrad. Stab.* **2002**, *77*, 227–233. [CrossRef]
7. Price, D.; Cunliffe, L.K.; Bullett, K.J.; Hull, T.R.; Milnes, G.J.; Ebdon, J.R.; Hunt, B.J.; Joseph, P. Thermal behavior of covalently bonded phosphonate flame-retarded poly(methyl methacrylate) systems. *Polym. Adv. Technol.* **2008**, *19*, 710–723. [CrossRef]

8. Schartel, B.; Braun, U.; Balabanovich, A.I.; Artner, J.; Ciesielski, M.; Doring, M.; Perez, R.M.; Sandler, J.K.W.; Altstadt, V. Pyrolysis and fire behaviour of epoxy systems containing a novel 9,10-dihydro-9-oxa-10-phosphaphenanthrene-10-oxide-(DOPO)-based diamino hardener. *Eur. Polym. J.* **2008**, *44*, 704–715. [CrossRef]

9. Menard, R.; Negrell, C.; Ferry, L.; Sonnier, R.; David, G. Synthesis of biobased phosphorus-containing flame retardants for epoxy thermosets comparison of additive and reactive approaches. *Polym. Degrad. Stab.* **2015**, *120*, 300–312. [CrossRef]

10. Sonnier, R.; Ferry, L.; Lopez-Cuesta, J.M. Flame Retardancy of Phosphorus-Containing Polymers. In *Phosphorus-Based Polymers: From Synthesis to Applications*; Royal Society of Chemistry: London, UK, 2014.

11. Hergenrother, P.M.; Thompson, C.M.; Smith, J.G., Jr.; Connell, J.W.; Hinkley, J.A.; Lyon, R.E.; Moulton, R. Flame retardant aircraft epoxy resins containing phosphorus. *Polymer* **2005**, *46*, 5012–5024. [CrossRef]

12. Braun, U.; Balabanovich, A.I.; Schartel, B.; Knoll, U.; Artner, J.; Ciesielski, M.; Döring, M.; Perez, R.; Sandler, J.K.; Altstädt, V. Influence of the oxidation state of phosphorus on the decomposition and fire behaviour of flame-retarded epoxy resin composites. *Polymer* **2006**, *47*, 8495–8508. [CrossRef]

13. Lorenzetti, A.; Modesti, M.; Besco, S.; Hrelja, D.; Donadi, S. Influence of phosphorus valency on thermal behaviour of flame retarded polyurethane foams. *Polym. Degrad. Stab.* **2011**, *96*, 1455–1461. [CrossRef]

14. Mariappan, T.; Zhou, Y.; Hao, J.; Wilkie, C.A. Influence of oxidation state of phosphorus on the thermal and flammability of polyurea and epoxy resin. *Eur. Polym. J.* **2013**, *49*, 3171–3180. [CrossRef]

15. Rabe, S.; Chuenban, Y.; Schartel, B. Exploring the modes of action of phosphorus-based flame retardants in polymeric systems. *Materials* **2017**, *10*, 455. [CrossRef]

16. Lyon, R.E.; Speitel, L.; Walters, R.N.; Crowley, S. Fire-resistant elastomers. *Fire Mater.* **2003**, *27*, 195–208. [CrossRef]

17. Dumitrascu, A.; Howell, B.A. Flame retardant polymeric materials achieved by incorporation of styrene monomers containing both nitrogen and phosphorus. *Polym. Degrad. Stab.* **2012**, *97*, 2611–2618. [CrossRef]

18. Joseph, P.; Tretsiakova-McNally, S. Combustion behaviours of chemically modified polyacrylonitrile polymers containing phosphorylamino groups. *Polym. Degrad. Stab.* **2012**, *97*, 2531–2535. [CrossRef]

19. Deng, Y.; Wang, Y.Z.; Ban, D.M.; Liu, X.H.; Zhou, Q. Burning behavior and pyrolysis products of flame-retardant PET containing sulfur-containing aryl polyphosphonate. *J. Anal. Appl. Pyrolysis* **2006**, *76*, 198–202. [CrossRef]

20. Howell, B.; Daniel, Y. The impact of sulfur oxidation level on flame retardancy. *J. Fire Sci.* **2018**, *36*, 518–534. [CrossRef]

21. Chang, T.; Chen, Y.; Ho, S.; Chiu, Y. The effect of silicon and phosphorus on the degradation of poly (methyl methacrylate). *Polymer* **1996**, *37*, 2963–2968. [CrossRef]

22. Sponton, M.; Mercado, L.; Ronda, J.; Galia, M.; Cadiz, V. Preparation, thermal properties and flame retardancy of phosphorus-and silicon-containing epoxy resins. *Polym. Degrad. Stab.* **2008**, *93*, 2025–2031. [CrossRef]

23. Cochez, M.; Ferriol, M.; Weber, J.; Chaudron, P.; Oget, N.; Mieloszynski, J. Thermal degradation of methyl methacrylate polymers functionalized by phosphorus-containing molecules I. TGA/FT–IR experiments on polymers with the monomeric formula CH2C (CH3) C (O) OCHRP (O)(OC2H5) 2 (R= H,(CH2) 4CH3, C6H5Br, C10H7). *Polym. Degrad. Stab.* **2000**, *70*, 455–462. [CrossRef]

24. Troev, K.; Kisiova, T.; Grozeva, A.; Borisov, G. Phosphorus-and metal-containing poly (ethylene terephthalate)—1. Phosphorus-and calcium-containing polymer. *Eur. Polym. J.* **1993**, *29*, 1205–1209. [CrossRef]

25. Troev, K.; Kisiova, T.; Grozeva, A.; Borisov, G. Phosphorus-and metal-containing poly (ethylene terephthalate)—2. Phosphorus-, calcium-and chlorine-containing polymer. *Eur. Polym. J.* **1993**, *29*, 1211–1215. [CrossRef]

26. Troev, K.; Kisiova, T.; Grozeva, A.; Borisov, G. Synthesis of phosphorus-, zinc-, and chlorine-containing poly (ethylene terephthalate). *J. Appl. Polym. Sci.* **1993**, *49*, 777–784. [CrossRef]

27. Liu, Y.L. Flame-retardant epoxy resins from novel phosphorus-containing novolac. *Polymer* **2001**, *42*, 3445–3454. [CrossRef]

28. Liu, Y.L.; Tsai, S.H. Synthesis and properties of new organosoluble aromatic polyamides with cyclic bulky groups containing phosphorus. *Polymer* **2002**, *43*, 5757–5762. [CrossRef]

29. Sablong, R.; Duchateau, R.; Koning, C.E.; Pospiech, D.; Korwitz, A.; Komber, H.; Starke, S.; Häußler, L.; Jehnichen, D.; der Landwehr, M.A. Incorporation of a flame retardancy enhancing phosphorus-containing diol into poly (butylene terephthalate) via solid state polycondensation: A comparative study. *Polym. Degrad. Stab.* **2011**, *96*, 334–341. [CrossRef]

30. Petreus, O.; Avram, E.; Serbezeanu, D. Synthesis and characterization of phosphorus-containing polysulfone. *Polym. Eng. Sci.* **2010**, *50*, 48–56. [CrossRef]

31. Negrell, C.; Frénéhard, O.; Sonnier, R.; Dumazert, L.; Briffaud, T.; Flat, J.J. Self-extinguishing bio-based polyamides. *Polym. Degrad. Stab.* **2016**, *134*, 10–18. [CrossRef]

32. Chang, S.J.; Sheen, Y.C.; Chang, R.S.; Chang, F.C. The thermal degradation of phosphorus-containing copolyesters. *Polym. Degrad. Stab.* **1996**, *54*, 365–371. [CrossRef]

33. Liu, Y.L.; Hsiue, G.H.; Lan, C.W.; Chiu, Y.S. Flame-retardant polyurethanes from phosphorus-containing isocyanates. *J. Polym. Sci. Part A Polym. Chem.* **1997**, *35*, 1769–1780. [CrossRef]

34. Tibiletti, L.; Ferry, L.; Longuet, C.; Mas, A.; Robin, J.J.; Lopez-Cuesta, J.M. Thermal degradation and fire behavior of thermoset resins modified with phosphorus containing styrene. *Polym. Degrad. Stab.* **2012**, *97*, 2602–2610. [CrossRef]

35. Vahabi, H.; Ferry, L.; Longuet, C.; Sonnier, R.; Negrell-Guirao, C.; David, G.; Lopez-Cuesta, J.M. Theoretical and empirical approaches to understanding the effect of phosphonate groups on the thermal degradation for two chemically modified PMMA. *Eur. Polym. J.* **2012**, *48*, 604–612. [CrossRef]

36. Ménard, R.; Negrell, C.; Fache, M.; Ferry, L.; Sonnier, R.; David, G. From a bio-based phosphorus-containing epoxy monomer to fully bio-based flame-retardant thermosets. *RSC Adv.* **2015**, *5*, 70856–70867. [CrossRef]

37. Dumitrascu, A.; Howell, B.A. Flame-retarding vinyl polymers using phosphorus-functionalized styrene monomers. *Polym. Degrad. Stab.* **2011**, *96*, 342–349. [CrossRef]

38. Sonnier, R.; Otazaghine, B.; Iftene, F.; Negrell, C.; David, G.; Howell, B.A. Predicting the flammability of polymers from their chemical structure: An improved model based on group contributions. *Polymer* **2016**, *86*, 42–55. [CrossRef]

39. Sonnier, R.; Otazaghine, B.; Dumazert, L.; Ménard, R.; Viretto, A.; Dumas, L.; Bonnaud, L.; Dubois, P.; Safronava, N.; Walters, R. Prediction of thermosets flammability using a model based on group contributions. *Polymer* **2017**, *127*, 203–213. [CrossRef]

40. Van Krevelen, D.W. *Properties of Polymers*; Elsevier: Amsterdam, The Netherlands, 1990; pp. 627–653.

41. Lyon, R.E.; Takemori, M.T.; Safronava, N.; Stoliarov, S.I.; Walters, R.N. A molecular basis for polymer flammability. *Polymer* **2009**, *50*, 2608–2617. [CrossRef]

42. Walters, R.N.; Lyon, R.E. Molar group contributions to polymer flammability. *J. Appl. Polym. Sci.* **2003**, *87*, 548–563. [CrossRef]

43. Sonnier, R.; Negrell-Guirao, C.; Vahabi, H.; Otazaghine, B.; David, G.; Lopez-Cuesta, J.M. Relationships between the molecular structure and the flammability of polymers: Study of phosphonate functions using microscale combustion calorimeter. *Polymer* **2012**, *53*, 1258–1266. [CrossRef]

44. Bier, F.; Six, J.L.; Durand, A. DOPO-Based Phosphorus-Containing Methacrylic (Co)Polymers: Glass Transition Temperature Investigation. *Macromol. Mater. Eng.* **2019**, *304*, 1800645–1800655. [CrossRef]

45. Atherton, F.; Openshaw, H.; Todd, A. Studies on phosphorylation. 2. the reaction of dialkyl phosphites with polyhalogen compounds in presence of bases-a new method for the phosphorylation of amines. *J. Chem. Soc.* **1945**, 660–663. [CrossRef]

46. Atherton, F.; Todd, A. 129. Studies on phosphorylation. Part III. Further observations on the reaction of phosphites with polyhalogen compounds in presence of bases and its application to the phosphorylation of alcohols. *J. Chem. Soc.* **1947**, 674–678. [CrossRef]

47. Canniccioni, B. Synthèse et caractérisation de copolymères à blocs amphiphiles et doubles hydrophiles à base de phosphore par polymérisation radicalaire contrôlée RAFT. Ph.D. Thesis, Montpellier University 2, Montpellier, France, 2013.

48. Lyon, R.E.; Walters, R.N. Pyrolysis combustion flow calorimetry. *J. Anal. Appl. Pyrolysis* **2004**, *71*, 27–46. [CrossRef]

49. Walters, R.N.; Safronava, N.; Lyon, R.E. A microscale combustion calorimeter study of gas phase combustion of polymers. *Combust. Flame* **2015**, *162*, 855–863. [CrossRef]

50. Huggett, C. Estimation of rate of heat release by means of oxygen consumption measurements. *Fire Mater.* **1980**, *4*, 61–65. [CrossRef]

51. Correa, S.M. A review of NOx formation under gas-turbine combustion conditions. *Combust. Sci. Technol.* **1993**, *87*, 329–362. [CrossRef]

52. Walters, R.N.; Hackett, S.M.; Lyon, R.E. Heats of combustion of high temperature polymers. *Fire Mater.* **2000**, *24*, 245–252. [CrossRef]

53. Sonnier, R.; Otazaghine, B.; Ferry, L.; Lopez-Cuesta, J.M. Study of the combustion efficiency of polymers using a pyrolysis–combustion flow calorimeter. *Combust. Flame* **2013**, *160*, 2182–2193. [CrossRef]

54. Sonnier, R.; Dorez, G.; Vahabi, H.; Longuet, C.; Ferry, L. FTIR–PCFC coupling: A new method for studying the combustion of polymers. *Combust. Flame* **2014**, *161*, 1398–1407. [CrossRef]

55. Lewin, M. Synergism and catalysis in flame retardancy of polymers. *Polym. Adv. Technol.* **2001**, *12*, 215–222. [CrossRef]

Article

Natural Keratin and Coconut Fibres from Industrial Wastes in Flame Retarded Thermoplastic Starch Biocomposites

Sebastian Rabe [1], Guadalupe Sanchez-Olivares [2], Ricardo Pérez-Chávez [2] and Bernhard Schartel [1,*

[1] Bundesanstalt für Materialforschung und-prüfung (BAM), 12205 Berlin, Germany; bernhard.schartel@bam.de

[2] CIATEC, A.C. Center of Applied Innovation in Competitive Technologies, Guanajuato 37545, Mexico; gsanchez@ciatec.mx (G.S.-O.); rperez@ciatec.mx (R.P.-C.)

* Correspondence: bernhard.schartel@bam.de; Tel.: +49-30-81041021

Received: 14 December 2018; Accepted: 15 January 2019; Published: 22 January 2019

Abstract: Natural keratin fibres derived from Mexican tannery waste and coconut fibres from coconut processing waste were used as fillers in commercially available, biodegradable thermoplastic starch-polyester blend to obtain sustainable biocomposites. The morphology, rheological and mechanical properties as well as pyrolysis, flammability and forced flaming combustion behaviour of those biocomposites were investigated. In order to open up new application areas for these kinds of biocomposites, ammonium polyphosphate (APP) was added as a flame retardant. Extensive flammability and cone calorimeter studies revealed a good flame retardance effect with natural fibres alone and improved effectiveness with the addition of APP. In fact, it was shown that replacing 20 of 30 wt. % of APP with keratin fibres achieved the same effectiveness. In the case of coconut fibres, a synergistic effect led to an even lower heat release rate and total heat evolved due to reinforced char residue. This was confirmed via scanning electron microscopy of the char structure. All in all, these results constitute a good approach towards sustainable and biodegradable fibre reinforced biocomposites with improved flame retardant properties.

Keywords: biomaterials; biodegradable; calorimetry; composites; flame retardance

---

1. Introduction

The recycling of industrial wastes is an economically interesting and environmentally friendly approach towards sustainable material resources. For an efficient manufacturing and usage lifecycle, it is necessary for the processing plant to be in the vicinity of the material source. The natural fibre wastes investigated in this paper are derived directly from Mexican industrial sectors and processed in Mexico, to reduce transport and storage effort to a minimum.

In order to create a completely biodegradable biocomposite obtained from renewable resources, synthetic fillers such as carbon or glass fibres need to be replaced and a biopolymer must be used as matrix [1–3]. An inexpensive and ecological and thus sustainable, alternative is found in natural fibres derived from industrial process wastes. These may come from plant-based sources like bast fibres as well as animal sources [4]. Vegetal fibres are composed of cellulose, hemicellulose and lignin, while animal-based fibres consist mainly of proteins, posing a valuable nitrogen source. In polymer composites, the main application of natural fibres has been as mechanical reinforcement [5]. The properties of the natural fibre reinforced polymer composites depend on different variables including fibre type, aspect ratio (length/width), content, modification method, interaction with the polymer matrix and processing conditions [6]. For example, sisal fibre lengths greater than 10 mm (20 and 30 mm) improve the mechanical properties of sisal/polypropylene composites. The samples were obtained by hot pressing and the composites with 30% mass fraction of fibre exhibited tensile, flexural and impact strength greater than polypropylene. The authors mentioned that, when the fibre mass fraction is added at low content, the fibres may act as filler leading to a reduction in the strength of the composite. In addition, they compered their results with sisal-polystyrene composites, in which the low tensile strength was attributed to the random orientation of the short fibres [7]. Similarly, the mechanical performance of abaca, jute and flax fibres on polylactic acid (PLA) biocomposites was evaluated using 30 wt. % fibre content. Jute fibres showed better tensile strength than abaca and flax fibres, while abaca fibres increased the Charpy notched impact strength (total amount of energy that a material is able to absorb) more than jute and flax fibres did. The later results point out that, the natural fibre type has a strong influence on the mechanical properties of the composites. There are several types of natural fibres, which differ regarding their chemical structure [8]. Otherwise, reinforced PLA/abaca fibre biocomposites, obtained by extrusion and injection moulding with 30 wt. % abaca fibre content, demonstrated a tensile modulus improved by a factor of 2.4, flexural modulus improved by a factor of 1.20 and impact strength 2.4 times greater than unreinforced PLA. According to the morphology of the composites, abaca fibres seem to be well coated with PLA matrix ascribed to the surface roughness of abaca fibre. The author suggested that reinforcing PLA with natural fibres leads to good chemical bonding on the interphase [9].

Regarding animal-based fibres, they are characterized by different properties than vegetal fibres. For instance, wool keratin fibres possess surface toughness, flexibility and a high aspect ratio and are less hydrophilic than vegetal fibres. Feather keratin fibres have a hollow structure and thus present extremely low fibre density and a low dielectric constant (k). The potential of chicken feather keratin fibres has been used to produce a low-k dielectric composite for electronic applications. The composites were obtained using acrylated epoxidized soybean oil and keratin feather fibres. The hollow keratin fibres were not filled by resin infusion and the composite retained a significant volume of air in the hollow structure of the fibres. The resulting dielectric constant (k) of the composites depend on the keratin feather fibre volume fraction due to the retained air. The values obtained in this work were significantly lower than the conventional silicon dioxide or epoxy or polymer dielectric insulators [10]. In addition, feather keratin fibres have been used as a reinforcement agent in polymer composites. As an example, 20 wt. % of feather keratin fibres were added to high density polyethylene (HDPE). Temperature, time and speed during compounding step were studied. The authors reported the optimal conditions of the processing to maximize fibre-polymer interactions and minimize degradation of feather keratin fibres. The composite showed improved stiffness and lower density than virgin HDPE [11].

Thermoplastic starch is derived from potatoes, corn or other cereals and consists mainly of amylose and amylopectin. Due to its high sensitivity towards hydrolysis as well as its poor dimensional stability and mechanical properties, the starch phase is blended with a biodegradable polyester [12]. This upgrades thermoplastic starch into a biodegradable material interesting for various applications. Commercially available starch blends are typically used as food packaging material or in disposable and compostable items and films [13]. The increased demand for biodegradable materials makes it necessary to research further potential application areas, for example in the transport, construction, electric and electronic industries. For the utilization of natural fibre composites in these sectors it is essential to assess their fire retardancy [14]. Thermoplastic starch has demonstrated good flame retardancy effects through the combination of coconut fibres and aluminium trihydroxide; the heat release rate, fire growth rate and total fuel release were reduced significantly. These results were explained by the increase in carbonaceous char accompanied by the reduction of carbon content in the pyrolysis products. Moreover, using coconut fibres on thermoplastic starch allowed the reduction of the aluminium trihydroxide content. The results offer a foundation for reducing the content of a traditional flame retardant additive by adding natural fibres to thermoplastic starch biocomposites [15]. The thermoplastic starch- blends, one of which was used as a polymer matrix in this publication, were shown to decompose to a satisfying degree in a composting environment and generally achieve major degradation rating in environmental exposure [16] Degradation in anaerobic conditions showed a decomposition similar to cellulose filter paper, measured by decrease of volatile solids [17]. In general, these thermoplastic starch-blend products are made with regards to compostability in order to reduce organic waste landfill disposal and thus biogas production, the main hazard of organic waste.

Flame retardants based on phosphorus are widely used to replace halogen-based additives, which exhibit toxicity and pose environmental risks [18–20]. Several flame retardants based on phosphorus are available on the market [21–24]. Phosphate compounds have been used extensively in order to reduce the flammability of cellulosic materials. For example, the effect of ammonium polyphosphate (APP) as a flame retardant additive, in combination with natural fibres, has been investigated for various polymer matrices. Polypropylene, polyurethane and thermoplastic starch using wood flake and corn shell in combination with APP exhibited flame retardancy improved up to the self-extinguishing V0 rating. From this research, it was observed a charring behaviour of polysaccharides and polyurethane in presence of APP. These results demonstrated the efficiency of APP as a flame retardant additive for biopolymer systems [25].

The present work focuses on the processability and properties of natural fibre biocomposites with an emphasis on flame retardant behaviour with and without APP as a flame retardant. The synergism between natural fibres and APP in flammability and burning behaviour under forced flaming conditions is exposed and explained. One of the main objectives of this work is to reduce the high content of APP added to conventional flame retardant composites by taking advantage of waste materials but maintaining or even improving the mechanical and flame retardant properties of the conventional composites. Moreover, to the best of our knowledge, the use of natural keratin and coconut fibres derived from industrial wastes in combination with halogen free ammonium polyphosphate to produce fire retardant polymer composites based on biodegradable thermoplastic starch has not been reported previously.

2. Materials and Methods

Thermoplastic starch (TPS), Mater-Bi® EF05B, 100% compostable biopolymer, was purchased from NOVAMONT SpA (Novara, Italy). Keratin fibres (KF) were recovered as waste from the beamhouse stage of a Mexican tannery. Coconut fibres (CF) were acquired from COPEMASA Co. (Tecomán, Mexico) as waste from the husks of coconut fruits. Ammonium polyphosphate (APP), trade name Exolit AP-422, was supplied by Clariant GmbH (Wiesbaden, Germany), with 31–32 wt. % phosphorus content and a particle size of D50 < 17 μm.

The KF were obtained as a by-product of waste hair from the beamhouse stage after liming during processing at a tannery. KF treatment was carried out using a tannery test drum. The process consists of two main steps, deliming and degreasing. For deliming, ammonium sulphate and sodium bisulphite were added at a concentration of 0.5% with respect to the weight of the waste hair. The drum was rolled for 2 h after rinsing. The process was repeated until a neutral pH was achieved (7.5–8.5). The degreasing step was similar but used a tensoactive product at 1% concentration with respect to waste hair weight. Then the KF were drained and dried in an oven at 75 °C for 12 h. Finally, KF were sieved through a 2 mm mesh.

The CF were pulverized twice in a 7.5 HP ASF P200 machine from Alimentos y Servicios Funcionales, S.A. de C.V. (Mexico City, Mexico), first through a 5 mm mesh and then through a 1 mm mesh. Afterward the pulverized coconut fibres were sieved through a #60 mesh (250 μm). Before the compounding step, the CF were dried at 80 °C for 12 h.

TPS composites were obtained by extrusion in a twin-screw Leistritz Micro 27 extruder from Leistritz Advanced Technologies Corp. (Nuremberg, Germany), L/D = 32, with a diameter of 27.0 mm and 8 heating zones, using counterrotating intermeshing mode. The compounding was carried out under the following temperature profile: 130/135/135/140/145/150/145/140 °C (from feed to die). In order to improve the dispersion of fibres and flame retardant additives, the composites were extruded twice at a rotational speed of 120 rpm. The TPS and composites were dried before extrusion at 105 °C in a model 30 low pressure dryer from Maguire Products Inc. (Des Moines, IA, USA), with 80 psi (0.5516 MPa) maintained during heating. After each extrusion step the composites were granulated in a 7.5 HP Paganí granulator (Pagani, Mexico City, Mexico) using a screen with a 5 mm mesh. Table 1 describes the composition of the investigated materials.

Table 1. Composition of investigated TPS-based biocomposites.

Composition	Acronym
Mater-Bi EF05B	TPS
TPS + 15% keratin fibres	TPS/15KF
TPS + 20% KF	TPS/20KF
TPS + 30% KF	TPS/30KF
TPS + 10% coconut fibres	TPS/10CF
TPS + 20% CF	TPS/20CF
TPS + 30% CF	TPS/30CF
TPS + 10% ammonium polyphosphate	TPS/10APP
TPS + 15% APP	TPS/15APP
TPS + 20% APP	TPS/20APP
TPS + 30% APP	TPS/30APP
TPS + 15% KF + 15% APP	TPS/15KF/15APP
TPS + 20% KF + 10% APP	TPS/20KF/10APP
TPS + 10% CF + 15% APP	TPS/10CF/15APP
TPS + 20% CF + 10% APP	TPS/20CF/10APP

Specimens for UL 94 testing, oxygen index (OI) determination, cone calorimeter and mechanical tests were obtained by injection moulding in a Milacron TM55 model machine from Milacron LLC, Karnataka, India, with 4 heating zones. The injection moulding conditions were used as follows: 145/150/155/155 °C temperature profile (from feed to die), 95 mm/s injection speed, 110 bar injection fill pressure and 45 s cooling time. Before injection moulding all TPS composites were dried under the same conditions as those used for extrusion.

Morphology analysis was performed in a JEOL JSM-7600F field emission scanning electron microscope (SEM, JEOL, Ltd., Akishima, Japan), using fractured samples of the composites. Elemental analysis was carried out by the incorporated energy dispersive spectroscopy (EDS) over the scanned area.

Mechanical tests were carried out using a tensile Instron 5565, from Instron (Norwood, MA, USA), machine at 50 mm/min crosshead speed following the ASTM D638 standard, with sample dimensions according to type I and a thickness of 3.1 ± 0.1 mm. Izod impact resistance was evaluated according to the Izod notched ASTM D256 standard, using samples 64 mm $\times$ 12.7 mm $\times$ 12.7 mm in size.

Rheological analysis was performed in a strain-controlled Ares-G2 rheometer, from TA-Instruments (New Castle, DE, USA), using parallel plates (25 mm diameter) at 165 °C and a gap of 1.0 mm under Small Amplitude Oscillatory Shear flow (SAOS); the tests were performed in a linear viscoelastic regime.

Thermal analysis of the samples was conducted on a Netzsch TG 209 F1 Iris (Selb, Germany). Portions of 5 mg of a specimen were pyrolyzed under nitrogen at a heating rate of 10 K/min.

Burning behaviour under forced flaming conditions was analysed in a Fire Testing Technologies cone calorimeter (East Grinstead, UK). The specimens (100 mm $\times$ 100 mm $\times$ 3.1 mm) were conditioned at 23 °C and 50% RH for at least 48 h prior to measurement, wrapped in an aluminium tray and measured at an external heat flux of 50 kW/m^2 and a distance of 35 mm from the cone heater, to avoid contact between the sample and the spark igniter or cone heater in case of strong intumescence. Heat flux impact uniformity on the specimens was shown to remain very similar up to a distance of 25 mm [26]. Samples were measured in duplicate unless the results deviated by more than 10%.

The flammability of the specimens was characterized by the limiting oxygen index (OI) and classification in the UL 94 test. Oxygen index measurements were conducted according to ISO 4589 with a sample size of 100 mm × 6.5 mm × 3.1 mm and UL 94 classifications were achieved according to IEC 60695-11-10 with a sample size of 127 mm × 12.7 mm × 3.1 mm. Specimens for flammability tests were conditioned similar to cone calorimeter specimens.

3. Results and Discussion

3.1. Morphology

The morphology of keratin and coconut fibres, APP additive and the composites was studied by Scanning Electron Microscopy (SEM). Figure 1 displays the morphology of keratin (Figure 1A) and coconut (Figure 1B) fibres, as well as APP (Figure 1C) and TPS (Figure 1D). According to Figure 1A, keratin fibre shows a cylindrical shape, with a diameter of approximately 80 μm; overlapped layers on the keratin fibre surface are clearly observed. Coconut fibre morphology (Figure 1B) presents a width of approximately 80 μm and consists of microfibres of different widths (approximately 10–15 μm) [27]. Figure 1C reveals APP morphology. According to the APP micrograph, agglomerates, irregularly shaped particles and different particle sizes (approximately 5–18 μm) are observed. Figure 1D displays the morphology of the fractured surface of the TPS matrix. The TPS micrograph shows well-defined fractured planes, characterized by thin layers oriented toward the direction of fracture. According to Figure 1D, TPS morphology exhibits a ductile fracture surface.

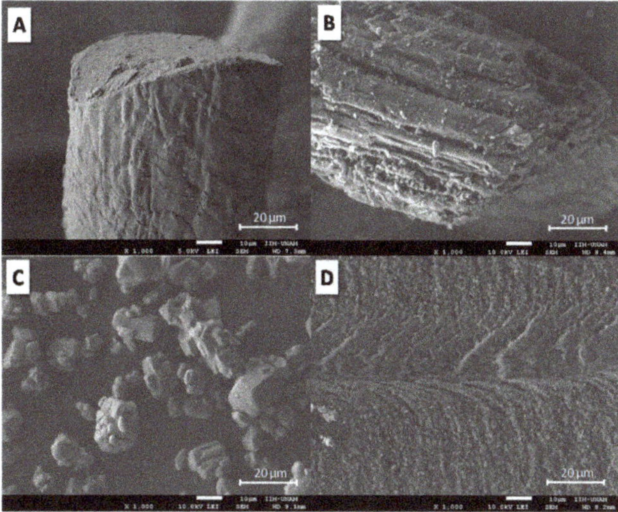

Figure 1. SEM micrographs of (**A**) KF, (**B**) CF, (**C**) APP and (**D**) TPS.

Figure 2 shows the morphology of fractured surfaces for TPP/20KF (Figure 2A,C) and TPS/20CF (Figure 2B,D) at two different magnifications. According to the SEM micrograph in Figure 2A,B, TPS loaded with 20 wt. % KF shows interstices around KF and empty holes caused by fibre extraction. Different fibre sizes were observed, with lengths of around 40–100 μm and diameters of approximately 30–50 μm. A similar morphology was observed for TPS/20CF composite in Figure 2C,D, where spaces around CF and footprints of fibre extraction are apparent. Coconut fibres showed a width of around 40–85 μm. According to Figure 2, the morphology of TPS/20KF and TPS/20CF suggests that fibres act as fillers in TPS without any relevant chemical or physical interactions.

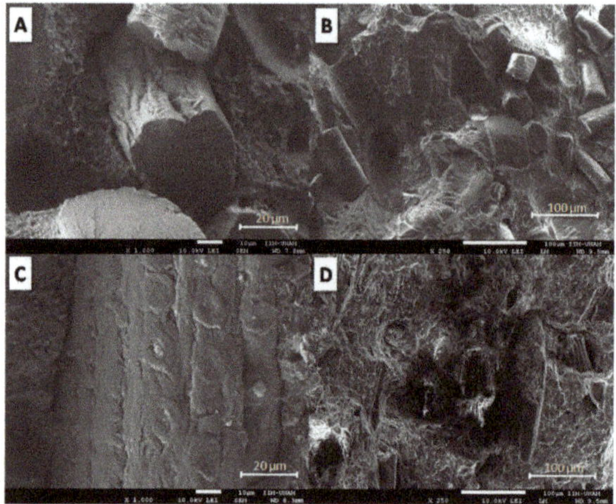

Figure 2. SEM micrographs of the fractured surface of TPS/20KF (**A,B**) and TPS/20CF (**C,D**) at two different magnifications.

Figure 3 shows SEM micrographs of the fractured surface of TPS using 20 wt. % APP. According to Figure 3A,B, APP additive exhibits poor adhesion with TPS, characterized by empty holes and interstices between the particles and the polymer. However, APP particles showed good dispersion in the polymer matrix; even though large particle sizes (around 20 μm) were observed, no APP agglomerates were visible. This morphology was achieved by the double extrusion process to obtain the well-dispersed TPS composites. In order to confirm the presence and dispersion of APP particles on the fractured surface of TPS, phosphorus elemental analysis was carried out by energy dispersive spectroscopy (EDS) of TPS/20APP composite (Figure 3C,D). Good dispersion of the phosphorus is shown by EDS analysis of the TPS/20APP fractured surface.

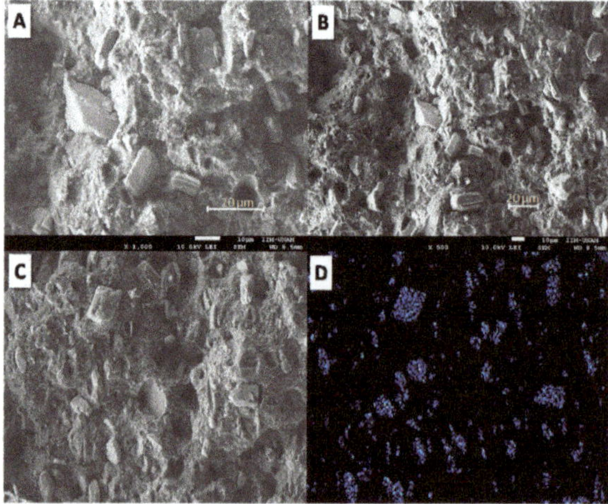

Figure 3. SEM micrographs of fractured surface of TPS/20APP at two different magnifications (**A,B**). Elemental analysis of TPS/20APP: (**C**) scanned area and (**D**) phosphorus mapping.

Figure 4 reveals the morphology of the fractured surface at different points and phosphorus elemental analysis of the TPS/20KF/10APP composite. According to Figure 4A,B, keratin fibres appear well embedded in the polymer matrix (indicated by black arrows); no interstices around the fibres were observed. With respect to APP particles, rather poor adhesion on the fractured surface was still apparent. Figure 4C,D shows the SEM image and the mapping of phosphorus in TPS/20KF/10APP; the micrographs reveal the presence and dispersion of APP particles. Figure 5 presents SEM micrographs and phosphorus elemental analysis of the fractured surfaces of the TPS/20CF/10APP composite. Figure 5A,B reveals dispersed coconut fibre and shows improved mechanical coupling around the fibre. Figure 5C,D illustrates the presence of APP particles with rather poor adhesion with the polymer matrix. According to morphological analyses of TPS/20KF/10APP and TPS/20CF/10APP composites (Figures 4 and 5 respectively), fibres added in combination with APP appeared better embedded on the polymer matrix than the composites without APP additive, probably because the higher load (30 wt. %) produces less polymer contraction in the composites.

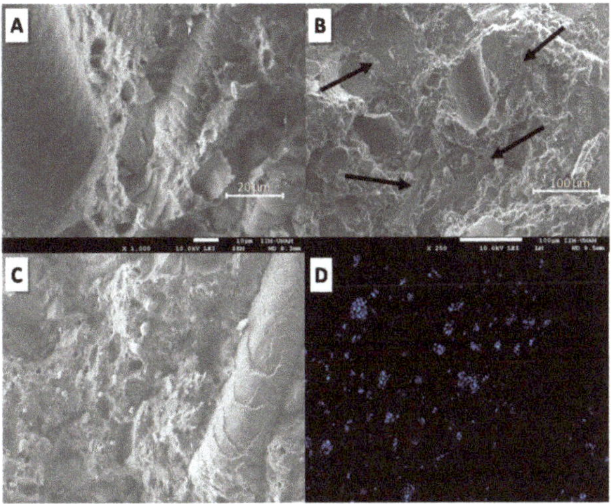

Figure 4. SEM micrographs of the fractured surface of TPS/20KF/10APP at two different magnifications (**A**,**B**). Elemental analysis of TPS/20KF/10APP: (**C**) scanned area and (**D**) phosphorus mapping.

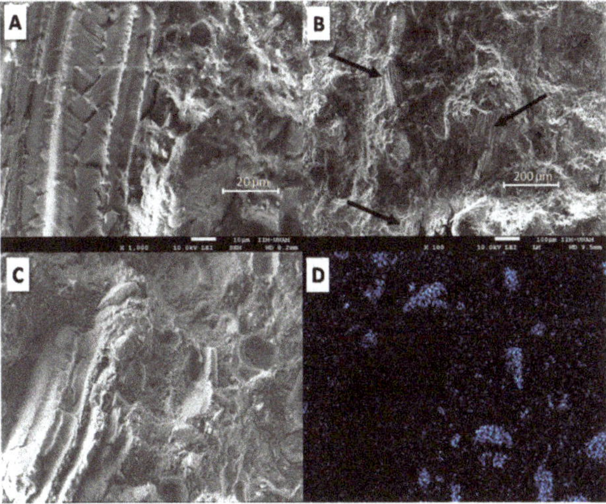

Figure 5. SEM micrographs of TPS/20CF/10APP at two different magnifications (**A**,**B**). Elemental analysis of TPS/20CF/10APP: (**C**) scanned area and (**D**) phosphorus mapping.

3.2. Mechanical Properties

Table 2 presents the mechanical properties of the investigated composites. According to Table 2, TPS shows ductile behaviour: a low Young's modulus (128 MPa), along with high strain at break (463%), tenacity (89 MPa) and Izod impact resistance (no break), as expected since this kind of biopolymer has traditionally been used for packaging applications [28]. Regarding the effect of keratin fibre contents on the mechanical properties of TPS composites, Table 2 reports that increasing KF content in TPS causes a reduction in strain at break, tenacity and Izod impact resistance. As an example, when KF was added at 15 and 30 wt. % (TPS/15KF and TPS/30KF), the strain at break was reduced from 463% to 190% and 26%, tenacity from 89 MPa to 19 and 2 MPa, Izod impact resistance from no break to 397 and 81 J/m, respectively. The Young's modulus increased from 128 MPa to 177 and 229 MPa, respectively. Tensile strength exhibited similar results for TPS/15KF and TPS/30KF composites, 11 and 10 MPa, respectively. Comparable mechanical behaviour was observed when CF was added to TPS: high CF content resulted in a high Young's modulus along with low strain at break, tenacity and Izod impact resistance. Taking a comparison of TPS, TPS/10CF and TPS/30CF as an example, strain at break was reduced from 463% to 207 and 13%, tenacity from 89 MPa to 31 and 2 MPa, Izod impact resistance from no break to 389 and 81 J/m, each in that order. Tensile strength presented the same value, 18 MPa, for both composites (TPS/10CF and TPS/30CF). The Young's modulus exhibited a remarkable increase, from 128 MPa to 168 and to 474 MPa, respectively, due to the high rigidity of coconut fibre [29]. According to the results on mechanical properties (Table 2), the increasing fibre content (KF and CF) in the TPS produced rigid and fragile materials, as already observed in natural fibre polymer composites [30]. A compatibilizer may be used in the future to obtain better mechanical coupling.

Table 2. Mechanical properties of composites.

Acronym	Young's Modulus $\pm \sigma$ (MPa)	Tensile Strength $\pm \sigma$ (MPa)	Strain at Break $\pm \sigma$ (%)	Tenacity $\pm \sigma$ (MPa)	Izod Impact Resistance $\pm \sigma$ (J/m)
TPS	128 ± 6	23 ± 1	463 ± 16	89 ± 8	non-break
TPS/15KF	177 ± 10	11 ± 0.3	190 ± 16	19 ± 2	397 ± 32
TPS/20KF	165 ± 13	11 ± 0.4	137 ± 23	13 ± 3	137 ± 0
TPS/30KF	229 ± 17	10 ± 0.1	26 ± 10	2 ± 1	81 ± 14
TPS/10CF	168 ± 6	18 ± 1	207 ± 4	31 ± 2	389 ± 26
TPS/20CF	322 ± 8	16 ± 1	40 ± 4	5 ± 1	230 ± 22
TPS/30CF	474 ± 25	18 ± 1	13 ± 2	2 ± 0.3	81 ± 20
TPS/10APP	124 ± 7	18 ± 0.5	368 ± 9	57 ± 2	non-break
TPS/15APP	138 ± 6	18 ± 0.3	336 ± 12	50 ± 2	non-break
TPS/20APP	163 ± 8	17 ± 0.3	233 ± 11	34 ± 4	213 ± 30
TPS/30APP	288 ± 11	15 ± 0.3	176 ± 7	24 ± 1	142 ± 21
TPS/15KF/15APP	247 ± 5	12 ± 1	54 ± 7	6 ± 1	118 ± 8
TPS/20KF/10APP	243 ± 3	13 ± 0.1	45 ± 3	5 ± 0.4	128 ± 8
TPS/10CF/15APP	225 ± 2	13 ± 0.1	61 ± 3	8 ± 0.1	147 ± 8
TPS/20CF/10APP	387 ± 5	14 ± 0.3	19 ± 5	2 ± 1	95 ± 14

As is well known, the addition of high content of flame retardants to polymer matrix may have a significant effect on mechanical properties [31]. The addition of 20 and 30 wt. % APP to TPS increased brittleness, as seen in TPS/10APP and TPS/15APP. Similarly, the TPS/20APP composite showed lower tensile strength, strain at break, tenacity and Izod impact resistance than TPS. The Young's modulus exhibited an increase from 128 to 163 MPa for TPS and TPS/20APP, respectively. The reinforcement achieved by APP is similar to that for CF.

Regarding KF in combination with the APP effect, Table 2 reports on TPS/15KF/15APP and TPS/20KF/10APP composites with different KF and APP contents but the same total load (30 wt. %), compounded in order to replace part of the APP flame retardant additive (5 or 10 wt. %) using KF. The resulting properties exhibited a similar Young's modulus, tensile strength and tenacity. Nevertheless, strain at break and Izod impact resistance were somewhat dependent on the composition of TPS/KF/APP: TPS/15KF/15APP exhibited 54% strain at break and 118 J/m Izod impact resistance, while TPS/20KF/10APP showed 45% strain at break and 128 J/m Izod impact resistance. Regarding CF in combination with APP (TPS/10CF/15APP and TPS/20CF/10APP composites), the mechanical properties were influenced by the CF content, as was observed in TPS/CF composites; high CF content reduced strain at break, tenacity and Izod impact resistance, while the Young's modulus was increased.

Nevertheless, although high KF and CF content, namely 30 wt. %, produces deteriorated mechanical properties, when KF and CF fibres were added at 20 wt. % in combination with 10 wt. % APP (30 wt. % total load), the corresponding mechanical properties were slightly improved over those achieved by adding only KF or CF at 30 wt. % (TPS/30KF and TPS/30CF). The combination of 20 wt. % KF and 10 wt. % APP (TPS/20KF/10APP) produced a less fragile and more ductile material than TPS/30KF. The Young's modulus, tensile strength, strain at break, tenacity and Izod impact resistance increased from 229 to 243 MPa, 10 to 13 MPa, 26 to 45%, 2 to 5 MPa and 81 to 128 J/m, respectively. This finding was similar to the combination of 20 wt. % CF with 10 wt. % APP (TPS/20CF/10APP) as compared to the TPS/30CF composite; strain at break and Izod impact resistance were slightly improved. In this context, the mechanical improvement in TPS/20KF/10APP and TPS/20CF/10APP composites over TPS/30KF and TPS/30CF indicates better stress transfer due to some physical fibre-polymer interactions [32], as was observed in morphological analyses (the fibres appeared well embedded in the polymer matrix, Figures 4 and 5).

3.3. Rheology

The rheological properties of TPS composites were assessed by oscillatory tests under continuous simple and small amplitude oscillatory shear flow (SAOS). Figure 6 displays complex viscosity (Figure 6A) and storage modulus (Figure 6B) as a function of angular frequency for TPS with different contents of 10, 15, 20 and 30 wt. % APP (TPS, TPS/10APP, TPS/15APP, TPS/20APP and TPS/30APP). According to the data plotted in Figure 6A, the complex viscosity of TPS and TPS/APP composites exhibited remarkable dependency on angular frequency, with complex viscosity decreasing as angular frequency increased. This behaviour is similar to the shear thinning effect observed in shear flow tests (not shown in this work). On the other hand, high APP content (20 and 30 wt. %) exerted an influence on complex viscosity throughout the entire frequency range (0.1–100 rad/s) greater than that for TPS. Regarding elastic properties, Figure 6B shows the storage modulus as a function of the angular frequency of TPS and TPS/APP composites (at 10, 15, 20 and 30 wt. %): the storage modulus exhibited higher values for samples using 20 and 30 wt. % APP than for TPS over the entire frequency range. The highest content of APP (TPS/30APP) produced the most elastic material, observed in the low frequency range (0.1–1.0 rad/s).

Figure 7 presents the effect of KF content and KF/APP combination on flow behaviour; complex viscosity (Figure 7A) and storage modulus (Figure 7B) are plotted as a function of angular frequency. Regarding the effect of KF content, Figure 7A shows that TPS/KF composites exhibited a rheological pattern similar to TPS; in the low frequency range (0.01–0.1 rad/s) the TPS/15KF composite presented curves overlapping with TPS, whereas in the high frequency range (1.0–100 rad/s), complex viscosity was slightly lower than for TPS. Meanwhile, TPS/20KF and TPS/30KF viscosity was slightly higher than TPS over the entire frequency range. In general, for all TPS/KF composites as well as TPS, complex viscosity showed a dependency on angular frequency similar to shear thinning behaviour in simple shear flow. The shear thinning effect was observed for all TPS/KF composites regardless of KF content. In this regard, it is reasonable to think that keratin fibres are well oriented under flow, which is consistent with similar natural fibre composites [33,34]. Figure 7B reveals that TPS/15KF and TPS have a similar storage modulus, with the TPS/15KF and TPS curves overlapping in the 0.1–1.0 rad/s range, while the storage modulus values for TPS/20KF and TPS/30KF were slightly higher than TPS over the entire frequency range. It is interesting to observe that the rheological behaviour of composites using keratin fibres depended mainly on the composition of the materials. Figure 7 shows that the KF and APP combination has a remarkably stronger effect on the complex viscosity and storage modulus than KF alone. When KF and APP were added to TPS (TPS/15KF/15APP and

TPS/20KF/10APP), complex viscosity and storage modulus increased over those of TPS and TPS/KF composites throughout the entire frequency range. However, rheological analyses point out that when KF are added in combination with APP, processing becomes more difficult than for KF without APP.

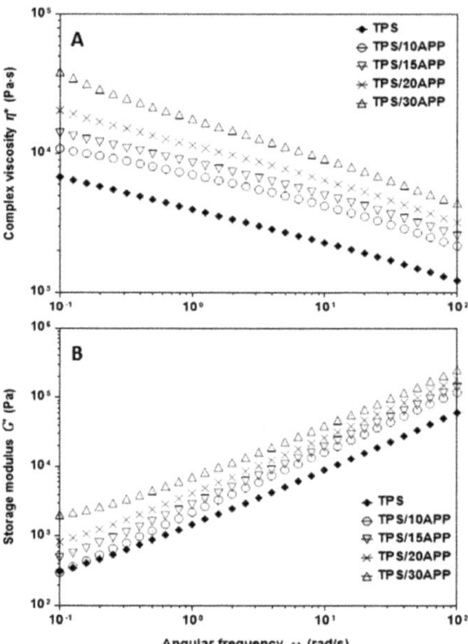

Figure 6. Complex viscosity (**A**) and storage modulus (**B**) as a function of the oscillatory frequency of TPS and TPS/APP composites with varied contents.

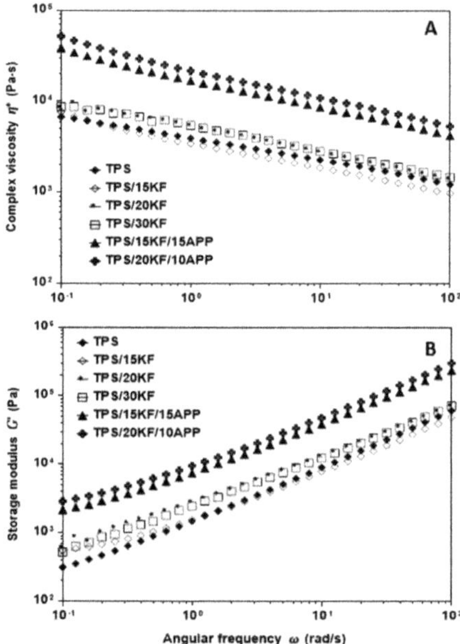

Figure 7. Complex viscosity (**A**) and storage modulus (**B**) as a function of the oscillatory frequency of TPS, TPS/KF and TPS/KF/APP composites with varied content.

Figure 8 displays the rheological behaviour of TPS using different contents of CF and the CF/APP combination. As it was observed for TPS, TPS/APP and TPS/KF composites, all composites using coconut fibre (Figure 8A) showed shear thinning behaviour (complex viscosity was reduced as frequency increased). On the other hand, CF content exhibited an important effect on complex viscosity; high CF content (20 and 30 wt. %) produced much higher complex viscosity than TPS, TPS/10CF, TPS/10CF/15APP and TPS/20CF/10APP, suggesting that CF behaved as a filler without physical or chemical interaction with the polymer matrix. Regarding storage modulus, Figure 8B shows an increase for TPS/10CF, TPS/10CF/15APP and TPS/20CF/10APP composites with reference to TPS but a storage modulus pattern similar to TPS (similar slopes). TPS/20CF and TPS/30CF composites exhibited much higher elastic properties than TPS/10CF, TPS/10CF/15APP and TPS/20CF/10APP composites. In this regard, it is possible to point out that the combination of coconut fibres with APP appears to favour the processing of these materials.

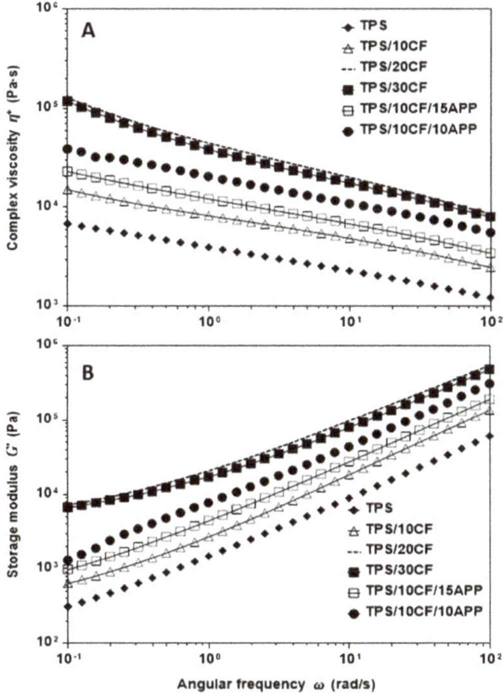

Figure 8. Complex viscosity (**A**) and storage modulus (**B**) as a function of the oscillatory frequency of TPS, TPS/CF and TPS/CF/APP composites with varied content.

3.4. Thermal Decomposition

Thermal decomposition under nitrogen was investigated using thermogravimetric analysis. TPS as well as the formulations with CF, KF, APP and their combinations were analysed to make a statement about altered decomposition pathways. Results of the thermogravimetry are displayed in Table 3. The temperature at 5 wt. % mass loss (Tat 5% Mass Loss) characterizes the beginning of the decomposition, the temperature at the maximum mass loss rates (Tmax) and the mass loss step (Δmass) the decomposition steps.

Table 3. Thermogravimetry results for all formulations.

Materials	Tat 5% Mass Loss (°C)	Tmax1 (°C)	Tmax2 (°C)	Δmass1 (%)	Δmass2 (%)	Residue at 800 °C (wt. %)
TPS	288	304	402	22	74	4
TPS/15KF	238	322	397	25	67	7
TPS/20KF	250	326	395	27	62	9
TPS/30KF	244	325	390	31	54	13
TPS/10CF	258	312	399	31	61	6
TPS/20CF	250	313	398	33	58	7
TPS/30CF	260	313	397	36	50	13
TPS/10APP	229	230	370	11	70	18
TPS/15APP	228	227	358	10	66	22
TPS/20APP	217	225	359	9	68	23
TPS/30APP	225	225	363	9	70	20
TPS/15KF/15APP	211	223	361	9	63	27
TPS/20KF/10APP	214	229	383	11	66	22
TPS/10CF/15APP	213	220	360	10	64	25
TPS/20CF/10APP	206	215	384	10	64	24

TPS shows two main decomposition steps, which are clearly visible in the mass loss rate (DTG) curve (Figure 9A). The first peak is attributed to the starch phase, with occurred at a temperature of 304 °C. The second decomposition step, at 402 °C, is associated with the polyester phase of the TPS. In the DTG, a shoulder is visible prior to the second peak, which derived from the additional additives incorporated in the commercial blend [35].

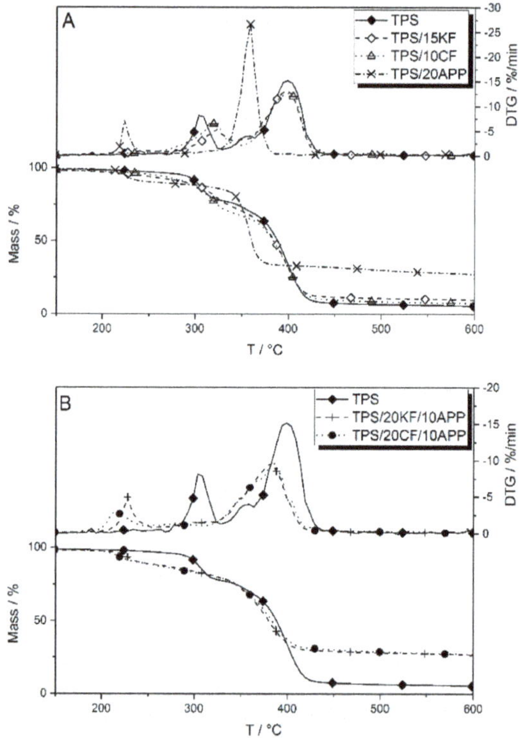

Figure 9. Mass and mass loss rates (DTG) of TPS blended with single natural fibre components or APP (**A**) and of combinations of TPS with natural fibres and APP (**B**).

Both fibres, KF and CF, increased the residue in thermogravimetry in a similar manner with increasing fibre content. The decomposition temperature of the first decomposition step was shifted towards higher temperatures, by 18 °C for 15 wt. % KF and by 22 °C for 10 wt. % CF in TPS, while the temperature for the second decomposition step remained more or less constant. The incorporation of natural fibres led to slightly reduced mass loss in both decomposition steps.

When APP was present in TPS formulations, the pyrolysis and decomposition temperatures changed significantly. With 20 wt. % of APP in TPS, the first decomposition step occurred at around 225 °C and was associated with the release of ammonia. A second decomposition step occurred at 359 °C, a temperature between the first and second steps of the reference TPS. This indicates an interaction between the TPS and the APP additive that led to an altered decomposition pathway during pyrolysis. The amount of residue formed by TPS/20APP is greatly increased, to 19 wt. % more than TPS.

Figure 9B presents mass and mass loss rates (DTG) curves for combinations of natural fibres and APP. They show the same first decomposition step as TPS/20APP, which is attributed to the loss of ammonia. This decomposition step occurred at similar temperatures when KF was incorporated into the APP-TPS blend and was shifted to lower temperatures when CF was used as a natural fibre filler. The second decomposition step was altered as compared to only APP in TPS, in a similar fashion for both fibre fillers. The maximum temperature of the second decomposition step was shifted to temperatures around 20 °C higher. Decomposition also happened at a slower pace and over a broader temperature range compared to only APP in TPS, as seen in a lower but broader peak in the DTG curves. This indicates an interaction between the fibre fillers and APP during thermal degradation. The combination of natural fibres and APP also led to increased residue formation. When comparing the additional residue of TPS/15KF/15APP, the measured residue of 27 wt. % was slightly more than the expected residue of 25 wt. %, suggesting slightly synergistic behaviour, while the formulation TPS/20KF/10APP achieved superposition in residue formation. The expected residue formation of around 24 wt. % for the combination TPS/10CF/15APP was similar to the measured residue amount of 25 wt. %. A good synergistic effect in residue formation is observed in TPS/20CF/10APP, with 24 wt. % compared to an expected amount of only 21 wt. %.

3.5. Flammability

The flammability of TPS biocomposites with the natural fibre fillers KF and CF, APP and the combination of both, were characterized by determining their limiting oxygen index and their classification in the UL 94 vertical test. UL 94 vertical test evaluates the reaction of the plastic materials to a small flame. The classification of this method depends on the flammability characteristics. For example, V0 classification means that the burning stops within 10 s for each individual specimen, after the first or second flame application to the specimen. No flaming drips are allowed. Meanwhile, V2 rating means that the burning stops within 30 s for each individual specimen, after the first or second flame application to the specimen. Flaming drips are allowed. The results of both flammability tests are summarized in Table 4.

Table 4. OI and UL 94 vertical test results of investigated biocomposites; no vertical rating (n. r.).

Formulation	OI (vol. %)	UL 94 Vertical
TPS	19.3	n. r.
TPS/15KF	20.3	n. r.
TPS/20KF	20.7	n. r.
TPS/30KF	21.1	n. r.
TPS/10CF	19.5	n. r.
TPS/20CF	20.3	n. r.
TPS/30CF	20.4	n. r.
TPS/10APP	24.4	n. r.
TPS/15APP	26.2	V-2
TPS/20APP	29.0	V-0
TPS/30APP	32.9	V-0
TPS/15KF/15APP	28.2	V-0
TPS/20KF/10APP	26.3	V-2
TPS/10CF/15APP	27.4	V-2
TPS/20CF/10APP	26.9	V-2

TPS as well as TPS-natural fibre biocomposites did not achieve a rating in the UL 94 vertical test. With increasing fibre content, the OI increased only slightly. However, a loading of 20 wt. % of APP in TPS resulted in an OI of 29 vol. % and was also the lowest concentration measured to achieve a V-0 classification in the UL 94 vertical test. It is remarkable that the minimum APP content to achieve a rating in the vertical test was lowered by the addition of keratin or coconut fibres. Even with 20 wt. % fibre content and only 10 wt. % APP content, a V-2 rating was achieved. OI values for all combinations of fibres and APP were in the range of 27 to 29 vol. %.

Figure 10 displays the change in OI behaviour with increasing additive contents, that is, fibre, APP and combinations of natural fibres and APP. In the graphs, the thin solid lines represent the linear dependency of the OI on the additive content of a respective single component in TPS, while the thick solid line shows the oxygen index behaviour if a combination of both additives, fibres and APP, were to result in a superposition. For the combinations of fibres and APP that amount to an overall additive content of 30 wt. %, namely TPS/15KF/15APP, TPS/20KF/10APP (Figure 10A) and TPS/20CF/10APP (Figure 10B), the measured oxygen indices are all located above the superposition line. This illustrates synergistic behaviour in OI. For CF, this synergism between fibres and APP was even more pronounced than for formulations with KF and APP, suggesting a better interaction between CF and APP during burning in terms of flame retardancy. These synergistic effects also explain the improved UL 94 rating for natural fibre-APP combinations.

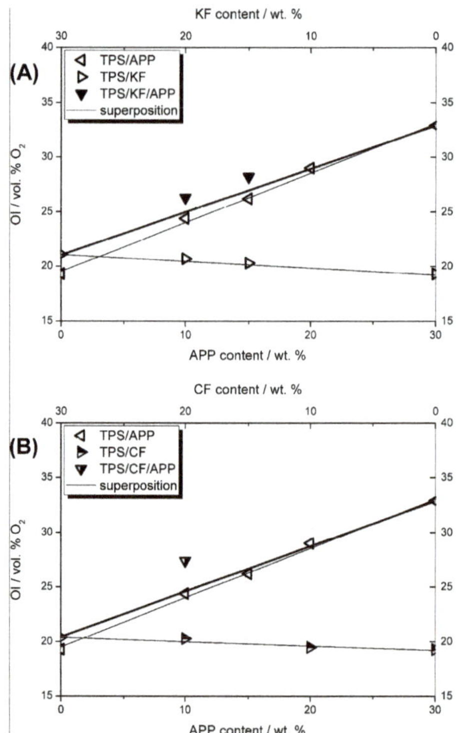

Figure 10. Content dependency and synergy of oxygen index for KF (**A**) and CF (**B**) formulations. Thin lines represent the linear behaviour of the OI for the single components derived from OI measurements of TPS/APP (hollow triangles pointing left), TPS/KF (hollow triangles pointing right) and TPS/CF (half-filled triangles pointing right) formulations, the thick line represents the calculated superposition of the fibre/APP combinations. The OI measured for fibre/APP combinations in TPS are displayed as solid triangles pointing down (TPS/KF/APP) and half-filled triangles pointing down (TPS/CF/APP).

The synergistic effect index was determined by calculating the theoretical superposition for linear behaviour using Equation (1) [36,37]. The OI for TPS/20CF and TPS/10APP increased compared to TPS by 1 vol. % and 5.1 vol. %, respectively. The calculated superpositioned OI increase for the formulation containing 20 wt. % of CF and 10 wt. % amounted to 6.1 vol. %. The oxygen index increase for TPS/20CF/10APP was measured to be 7.6 vol. %. Comparing the calculated superposition and the measured oxygen index increase yielded a synergistic effect index of 1.25.

$$SE_{abs}(OI)_{x,y=const.} = \frac{\Delta OI(TPS/20CF/10APP)}{\Delta OI(TPS/20CF) + \Delta OI(TPS/10APP)} = \frac{7.6 \text{ vol. \%}}{1 \text{ vol. \%} + 5.1 \text{ vol. \%}} = 1.25. \quad (1)$$

For TPS/20KF/10APP, the synergistic effect index, calculated in the same manner, amounted to 1.08. Through thorough observation of the OI flammability measurements, the synergistic effect, especially between CF and APP, is attributed to enhanced effectiveness in the condensed phase. This synergistic behaviour would also be expected in forced flaming combustion tests conducted in the cone calorimeter.

3.6. Burning Behaviour under Forced Flaming Combustion

The flame retarding effect of only keratin or coconut fibres in TPS is mainly limited to a PHRR (peak heat release rate) reduction of up to 33% and a THE (total heat evolved) reduction of up to 11%. Cone calorimeter results are shown in Table 5. The heat release rate (HRR) curves of natural fibres in TPS are shown in Figure 11.

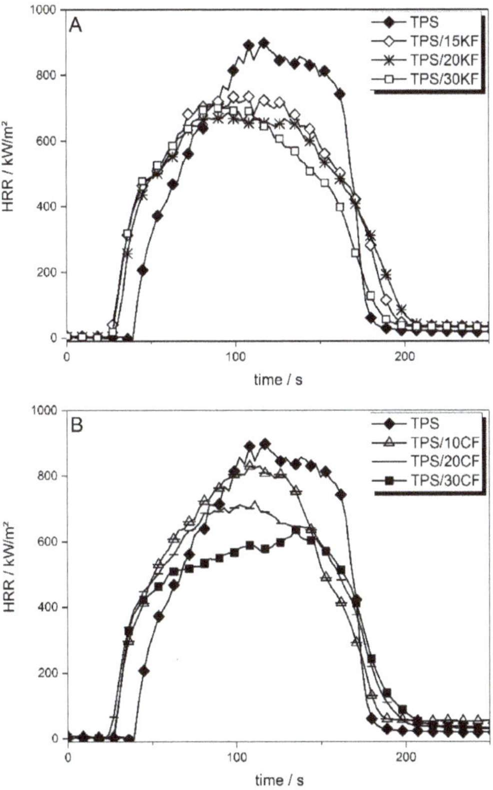

Figure 11. HRR curves for TPS reinforced with different amounts of keratin (**A**) and coconut fibres (**B**) in comparison to neat TPS.

Table 5. Cone calorimeter results of investigated TPS biocomposites.

Acronym	tig (s)	tfo (s)	PHRR (kW/m^2)	THE (MJ/m^2)	TML (g)	Residue (%)	THE/TML (MJ/gm^2)
TPS	37 ± 2	177 ± 3	960 ± 62	90.8 ± 0.7	44.0 ± 0.1	4.0 ± 0.0	2.1 ± 0.0
TPS/15KF	24 ± 2	193 ± 3	762 ± 1	90.1 ± 0.6	43.3 ± 0.2	8.0 ± 0.1	2.1 ± 0.0
TPS/20KF	28 ± 2	202 ± 7	690 ± 15	86.5 ± 4.2	42.1 ± 1.9	9.5 ± 0.2	2.1 ± 0.0
TPS/30KF	25 ± 2	192 ± 4	720 ± 2	80.6 ± 2.1	39.1 ± 0.5	10.7 ± 0.1	2.1 ± 0.0
TPS/10CF	24 ± 1	168 ± 19	864 ± 34	82.0 ± 7.8	38.8 ± 3.4	6.9 ± 0.2	2.1 ± 0.0
TPS/20CF	23 ± 2	199 ± 8	727 ± 3	87.8 ± 0.1	43.8 ± 0.1	9.9 ± 0.1	2.0 ± 0.0
TPS/30CF	26 ± 1	221 ± 3	644 ± 21	80.8 ± 4.2	43.4 ± 0.8	12.4 ± 0.1	1.9 ± 0.1
TPS/10APP	33 ± 0	235 ± 17	483 ± 14	52.6 ± 1.8	33.6 ± 1.3	23.3 ± 0.3	1.6 ± 0.0
TPS/15APP	33 ± 1	237 ± 17	438 ± 7	48.3 ± 1.1	32.2 ± 0.6	26.3 ± 0.0	1.5 ± 0.0
TPS/20APP	35 ± 1	261 ± 11	395 ± 2	44.0 ± 1.0	31.4 ± 0.3	28.5 ± 0.8	1.4 ± 0.0
TPS/30APP	43 ± 1	267 ± 7	376 ± 4	43.4 ± 0.7	29.6 ± 0.1	34.5 ± 0.2	1.5 ± 0.0
TPS/15KF/15APP	31 ± 1	271 ± 8	379 ± 5	49.9 ± 0.4	33.2 ± 1.5	24.3 ± 3.3	1.5 ± 0.1
TPS/20KF/10APP	31 ± 2	276 ± 12	407 ± 12	54.3 ± 1.6	33.6 ± 1.5	24.1 ± 0.3	1.6 ± 0.0
TPS/10CF/15APP	32 ± 2	240 ± 8	407 ± 1	42.2 ± 0.1	29.3 ± 0.4	30.0 ± 0.2	1.4 ± 0.0
TPS/20CF/10APP	27 ± 2	264 ± 2	338 ± 7	43.1 ± 1.1	29.6 ± 0.3	28.8 ± 0.1	1.5 ± 0.0

The incorporation of the natural fibres KF and CF had various effects on the HRR of TPS. For both sorts of fibres, the time to ignition (tig) was shifted to lower temperatures, from 37 s to around 25 s for 30 wt. % of fibres, respectively. Increased viscosity due to the addition of fibres resulted in later liquefaction of the sample surface in the cone calorimeter, reducing the heat exchange and thus cooling via convection. As a consequence, the sample heated up more quickly and ignited earlier. KF (Figure 11A) at 15 wt. % reduced the PHRR by around 200 kW/m^2. With higher amounts of KF, the reduction in PHRR levelled off. Apart from a slightly slower HRR decay after flameout, the overall burning behaviour was similar to non-flame retarded TPS. The effective heat of combustion, displayed here as total heat evolved (THE) divided by total mass loss (TML), was not influenced by the KF incorporated in TPS. They had an influence neither on the combustion efficiency of released fuel nor on the heat of combustion of the volatiles. When CF was incorporated into TPS, the levelling off of the reduction in PHRR was less pronounced, resulting in a lower PHRR for TPS/30CF (Figure 11B). The burning behaviour of TPS displayed a more characteristic change in the HRR curve with increasing amounts of CF than with added KF. For TPS/30CF, the PHRR was shifted to later times, just before flameout (tfo) and the fire growth behaviour in the beginning was reduced. This resulted in a decrease in the slope of the HRR curve after the initial rise. This indicates a more pronounced protective layer becoming visible in the HRR curve shape. THE/TML was only insignificantly decreased through the addition of CF. CF also produced a higher amount of residue, with 12.4 wt. % at a loading of 30 wt. % compared to 10.7 wt. % for TPS/30KF, hinting at higher thermal stability of the coconut fibres. In general, residue formation in the cone calorimeter was similar to thermogravimetric residue observation.

The incorporation of APP in TPS resulted in a clear change in HRR curve shape and thus in burning behaviour, as compared to that of a charring material [38]. The HRR curves of 10, 15, 20 and 30 wt. % APP are displayed in Figure 12.

HRR changed significantly for formulations with APP as compared to TPS. The initial peak in HRR after ignition, the ensuing local minimum and the slow decay of HRR, point to typical burning behaviour of material forming a protective layer. PHRR was decreased by around 50% with an APP load of just 10 wt. %. A loading of 20 wt. % APP lowered the PHRR by an additional 9%. Even higher APP loadings no longer significantly reduced the PHRR, indicating that the effectiveness of APP levels off. A similar trend is seen in the reduction of fire load (total heat evolved, THE). The THE was reduced by 42% at an APP loading of 10 wt. %. The highest APP load in TPS, which was 30 wt. %, resulted in a THE reduction of 52%. The addition of APP to TPS induced char formation, yielding 23 wt. % of residue at an APP load of 10 wt. %. A levelling off trend was also observed in residue formation with increasing APP load. Considering the initial residue amount of 4 wt. % for TPS, the addition of 10 wt. % APP yielded an additional residue formation of 19 wt. %. Incorporation of 30 wt. % APP in

TPS resulted in an additional residue amount of 31 wt. %. APP also led to a reduction of up to 30% in THE/TML. Otherwise released carbonaceous species which act as fuel for the flame were stored in the formed residue, reducing the heat of combustion. Additionally, APP released NH_3, which does not contribute to the heat production, diluting the flame and therefore reducing combustion efficiency.

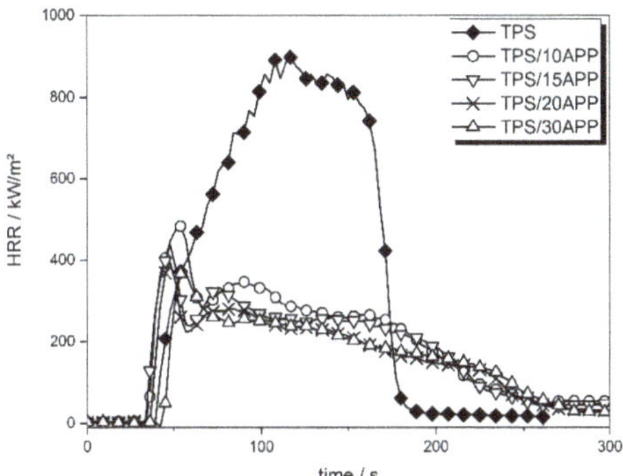

Figure 12. HRR curves of TPS formulations with APP.

Natural fibre reinforced TPS in combination with APP yielded similar HRR results as TPS formulations with only APP as a flame retardant additive. However, it is possible to replace certain amounts of the flame retardant APP with fibre content to achieve similarly good results as with 20 or 30 wt. % of APP alone. Combinations of natural fibre filler and APP are shown in Figure 13.

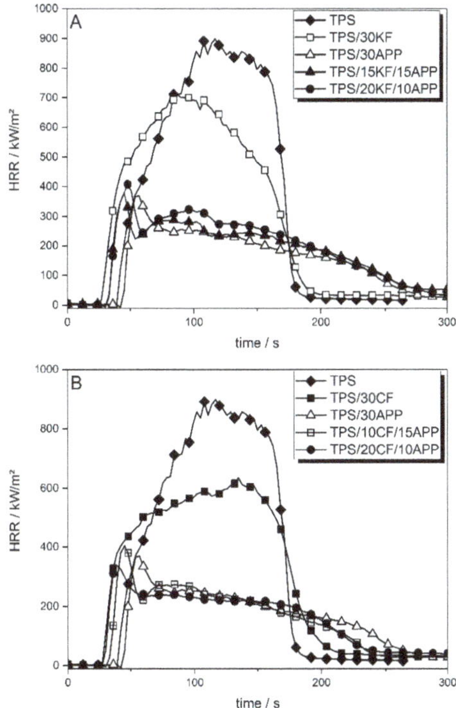

Figure 13. HRR curves for combinations of APP with keratin fibres (**A**) and coconut fibres (**B**).

Figure 13A displays the HRR curves derived from cone calorimeter measurements for a total additive amount of 30 wt. % in TPS. The KF-APP combinations TPS/15KF/15APP and TPS/20KF/10APP showed similar HRR results as TPS/30APP. TPS/15KF/15APP exhibited a reduction in PHRR to 379 kW/m^2 and a reduction in THE to 49.9 MJ/m^2, while TPS/20KF/10APP reduced the PHRR to 407 kW/m^2 and the THE to 54.3 MJ/m^2. The HRR was dominated by the addition of APP. The nonlinear levelling off of effectiveness with increasing APP content enabled strong synergy.

Combinations of CF and APP in TPS showed slightly different behaviour (Figure 13B). While the PHRR of TPS/10CF/15APP was still a bit higher than the PHRR of TPS/30APP, at 407 kW/m^2 compared to 376 kW/m^2, the THE was reduced to 42.2 MJ/m^2 as opposed to 43.4 MJ/m^2. A comparison of TPS/20CF/10APP and TPS/30APP shows that the overall flame retardancy performance may be even better even though a total additive load of 30 wt. % is maintained. The addition of 20 wt. % CF to the mixture of 10 wt. % APP in TPS resulted in a decrease in PHRR, which was not possible by increasing APP content alone. In terms of fire load, the THE of TPS/30APP and TPS/20CF/10APP were similarly low, at around 43.3 to 43.1 MJ/m^2.

For CF, a synergistic effect was observed in both PHRR and THE. In order to obtain the synergistic effect index as a statement for the significance of the synergism, Equation (2) is used [34], with M being the result derived from cone calorimeter measurement. This approach was chosen over the proportionate approach, in which the total additive load is kept constant, because there was no significant reduction in cone calorimeter results for TPS with 10, 20 or 30 wt. % APP. Since cone calorimeter results like PHRR or THE did not decrease in a linear fashion, the relative synergistic effect index was calculated as a more correct concept.

$$\text{SE}_{\text{rel}}(\text{M})_{x,y=\text{const.}} = \frac{1 - \Delta M(\text{TPS}/20\text{CF}/10\text{APP})}{1 - \Delta M(\text{TPS}/20\text{CF}) \times \Delta M(\text{TPS}/10\text{APP})}. \tag{2}$$

In TPS/20CF, the PHRR was reduced to 727 kW/m^2. APP at a load of 10 wt. % reduced the PHRR of TPS by around 50%. This resulted in an expected PHRR of TPS/20CF/10APP of around 366 kW/m^2, assuming no synergistic effect. However, the measured PHRR of TPS/20CF/10APP was 338 kW/m^2, yielding a synergistic effect index of 1.08. For THE, the synergism between CF and APP became even clearer. The calculated superposition in THE for TPS/20CF/10APP amounted to 50.9 MJ/m^2 and the measured THE was 43.1 MJ/m^2. Using Equation (2), this yielded a synergistic effect index of 1.2.

The effect of combining fibres and APP is visualized in Figure 14. The trapezoid schematics are constructed to illustrate the superposition values for 20 wt. % natural fibre and 10 wt. % APP in TPS. Hollow stars mark the superposition values of THE and PHRR for TPS/20KF/10APP (Figure 14A) and TPS/20CF/10APP (Figure 14B). For TPS/20KF/10APP, the measured values for THE and PHRR are located around the hollow star marking the calculated superposition. This illustrates the occurrence of a superposition effect of this combination. On the other hand, TPS/20CF/10APP showed clear synergistic behaviour in PHRR and especially in THE, since the measured values are located below the hollow star superposition.

Since APP and the natural fibre fillers showed their activity as flame retarding additives mostly in the condensed phase, the reason for the synergistic effect of CF in combination with APP is found by analysing the cone calorimeter residues.

Figure 15 shows the character of the cone calorimeter residues of TPS/10APP (Figure 15A) and the two fibre-APP combinations TPS/20KF/10APP (Figure 15B) and TPS/20CF/10APP (Figure 15C). When 10 wt. % of APP were incorporated in the TPS matrix, the formed residue was very light and brittle. The relatively smooth surface was disrupted by small cracks, resulting in a flaky and flocculent residue. Adding 20 wt. % of KF resulted in a more continuous surface but the overall character and nature of the residue was very similar to that of TPS/10APP. When 20 wt. % of CF were combined with 10 wt. % of APP, the char structure of the residue changed completely. Instead of a light and brittle char, the residue was very stable and showed a resemblance to wood char, pervaded by relatively large cracks. The change in char structure becomes clearer in the SEM micrographs of the respective residues.

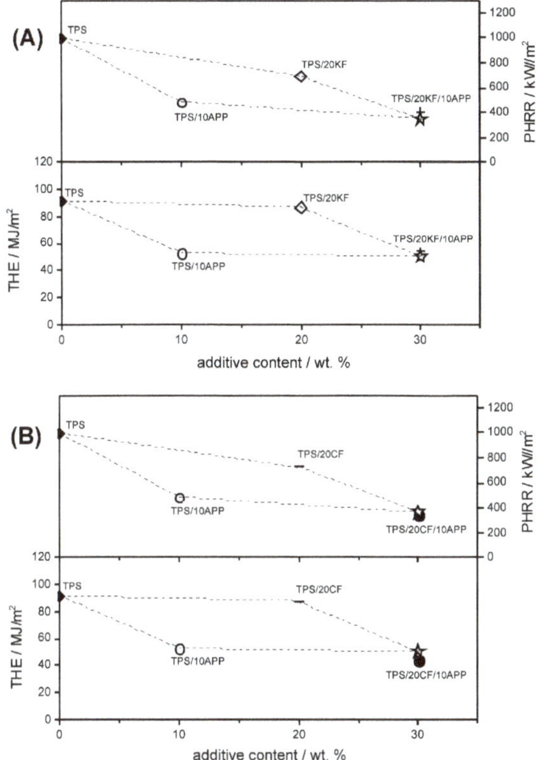

Figure 14. Visualization of superposition (hollow star) and measured PHRR and THE of TPS/20KF/10APP (**A**) and TPS/20CF/10APP (**B**).

Figure 15. Cone calorimeter residue photographs of TPS/10APP (**A**), TPS/20KF/10APP (**B**) and TPS/20CF/10APP (**C**).

Comparing the surfaces of TPS/20KF/10APP (Figure 16B) and TPS/10APP (Figure 16A) in SEM makes the similarities much clearer. Almost no differences are observed. It must be noted that no residual KF were found on the surface of the investigated TPS/20KF/10APP residue, so they do not contribute to the residue structure. In contrast to this, the residual CF were clearly visible in the TPS/20CF/10APP (Figure 16C) residue. CF reinforced the APP-induced char, resulting in a completely different char structure. This becomes even clearer at 400x magnification, where interconnection and coating of the CF with residue is apparent. The synergistic effect of CF in combination with APP in TPS is therefore attributed to the physical enhancement of the residue in TPS/20CF/10APP.

Figure 16. SEM micrographs of the cone calorimeter residue surfaces of TPS/10APP (**A**), TPS/20KF/10APP (**B**) and TPS/20CF/10APP (**C**) at two different magnifications (top and bottom).

4. Conclusions

With respect to mechanical properties, the addition of KF and CF to TPS has been demonstrated to produce rigid materials, as the Young's modulus was increased. However, tensile strength, strain at break, tenacity and Izod impact resistance were reduced. Comparing the effects of keratin and coconut fibre, CF produces a more rigid material than KF, namely at high content, 30 wt. %. On the other hand, the combination of KF or CF with APP (TPS/20KF/10APP or TPS/20CF/10APP) produces more ductile materials than those using 30 wt. % of KF or CF.

Regarding rheological properties, the addition of KF to TPS did not significantly affect the flow properties of TPS. However, when keratin fibres were added in combination with APP, complex viscosity increased, making processing more difficult. By contrast, when CF were added in combination with APP, the flow properties benefited and thus the processing of the composites also improves as compared with materials containing KF or CF without the flame retardant APP.

The flammability behaviour of TPS was greatly improved through the addition of APP. This was confirmed by both UL 94 and OI tests. In terms of the oxygen index, a synergistic effect was ascertained between APP and both fibres, KF and CF. For 20 wt. % KF and 10 wt. % APP in TPS, a synergistic effect index of 1.08 was calculated, while 20 wt. % CF in combination with 10 wt. % APP resulted in a synergistic effect index of 1.25. This indicated that CF improved the flame retardancy effectiveness of APP in TPS.

Investigating the fire behaviour under forced flaming conditions in the cone calorimeter revealed a change toward charring burning behaviour when the flame retardant APP was added. While the combination of KF and APP showed a good superposition, CF was able to further reduce PHRR and THE to values lower than those with APP as a single flame retardant additive. The synergistic effect of CF in combination with APP was quantified to an index of 1.08 in PHRR and 1.2 in THE reduction. This synergistic effect was attributed to the residue structure, since CF was able to reinforce the formed char, increasing residue stability. This was confirmed with SEM micrographs.

All in all, the investigation and characterization of the natural fibre fillers KF and CF for TPS enabled conclusions about mechanical and processing properties as well as great insights into flame retardancy behaviour. The combination of APP and industrial waste fibres is proposed as an interesting route towards sustainable flame retarded materials.

Author Contributions: Conceptualization, S.R., B.S. and G.S.-O.; validation, S.R. and R.P.-C.; investigation, S.R., R.P.-C. and G.S.-O.; resources, B.S. and G.S.-O.; data curation, S.R. and R.P.-C.; writing—original draft preparation, S.R. and G.S.-O.; writing—review and editing, B.S.; visualization, S.R. and G.S.-O.; supervision, G.S.-O. and B.S.; project administration, S.R., G.S.-O. and B.S.; funding acquisition, G.S.-O. and B.S.

Funding: This work was funded by the Bundesministerium für Bildung und Forschung (BMBF/01DN16040) and the Consejo Nacional de Ciencia y Tecnología (CONACYT/266441).

Acknowledgments: We would like to acknowledge Patrick Klack for his efforts in maintaining the apparatus, Jacqueline Gottwald for her help with flammability and pyrolysis measurements, Fausto Calderas for the training in rheological tests as well as Omar Novelo and Martin Günther for SEM analysis support.

Conflicts of Interest: The authors declare no conflict of interest.

References

1. Avérous, L.; Halley, P.J. Biocomposites based on plasticized starch. *Biofuel Bioprod. Biorefin.* **2009**, *3*, 329. [CrossRef]
2. Endres, H.-J.; Siebert-Raths, A. *Engineering Biopolymers*; Hanser Munich: Munich, Germany, 2008; p. 71, ISBN 978-3-446-42403-6.
3. Sypaseuth, F.D.; Gallo, E.; Çiftci, S.; Schartel, B. Polylactic acid biocomposites: Approaches to a completely green flame retarded polymer. *e-Polymers* **2017**, *17*, 449. [CrossRef]
4. Ramamoorthy, S.K.; Skrifvars, M.; Persson, A. A Review of Natural Fibers Used in Biocomposites: Plant, Animal and Regenerated Cellulose Fibers. *Polym. Rev.* **2015**, *55*, 107. [CrossRef]
5. Gallo, E.; Schartel, B.; Acierno, D.; Cimino, F.; Russo, P. Tailoring the flame retardant and mechanical performances of natural fiber-reinforced biopolymer by multi-component laminate. *Compos. Part B Eng.* **2013**, *44*, 112. [CrossRef]
6. Gurunathan, T.; Mohanty, S.; Nayak, S.K. A review of the recent developments in biocomposites based on natural fibres and their application perspectives. *Compos. Part A Appl. Sci. Manuf.* **2015**, *77*, 1. [CrossRef]
7. Jayaraman, K. Manufacturing sisal–polypropylene composites with minimum fibre degradation. *Compos. Sci. Technol.* **2003**, *63*, 367. [CrossRef]
8. Faruk, O.; Bledzki, A.K.; Fink, H.-P.; Sain, M. Biocomposites reinforced with natural fibers: 2000–2010. *Prog. Polym. Sci.* **2012**, *37*, 1552. [CrossRef]
9. Bledzki, A.K.; Jaszkiewicz, A.; Scherzer, D. Mechanical properties of PLA composites with man-made cellulose and abaca fibres. *Compos. Part A Appl. Sci. Manuf.* **2009**, *40*, 404. [CrossRef]
10. Zini, E.; Scandola, M. Green composites: An overview. *Polym. Compos.* **2011**, *32*, 1905. [CrossRef]
11. Barone, J.R.; Schmidt, W.F.; Liebner, C.F.E. Compounding and molding of polyethylene composites reinforced with keratin feather fiber. *Compos. Sci. Technol.* **2005**, *65*, 683. [CrossRef]
12. Balaji, A.B.; Pakalapati, H.; Khalis, M.; Walvekar, R.; Siddiqui, H. Natural and synthetic biocompatible and biodegradable polymers. In *Biodegradable and Biocompatible Polymer Composites Processing, Properties and Applications*; Shimpi, N.G., Ed.; Woodhead Publishing: Duxford, UK, 2018; pp. 14–15. ISBN 9780081009703.
13. Sreekumar, P.A.; Thomas, S. Matrices for natural-fibre reinforced composites. In *Properties and Performance of Natural-Fibre Composites*; Pickering, K.L., Ed.; Woodhead Publishing: Cambridge, UK, 2008; pp. 77–78. ISBN 9781845692674.
14. Galaska, M.L.; Horrocks, A.R.; Morgan, A.B. Flammability of natural plant and animal fibers: A heat release survey. *Fire Mater.* **2017**, *41*, 275. [CrossRef]
15. Gallo, E.; Sánchez-Olivares, G.; Schartel, B. Flame retardancy of starch—Based biocomposites—Aluminum hydroxide-coconut fiber synergy. *Polimery* **2013**, *58*, 395. [CrossRef]
16. de Wit, C.A. An overview of brominated flame retardants in the environment. *Chemosphere* **2002**, *46*, 583. [CrossRef]
17. Day, M.; Shaw, K.; Cooney, D.; Watts, J.; Harrigan, B. Degradable polymers: The role of the degradation environment. *J. Environ. Polym. Degrad.* **1997**, *5*, 137. [CrossRef]
18. Mohee, R.; Unmar, G.D.; Mudhoo, A.; Khadoo, P. Biodegradability of biodegradable/degradable plastic materials under aerobic and anaerobic conditions. *Waste Manag.* **2008**, *28*, 1624. [CrossRef]
19. Covaci, A.; Harrad, S.; Abdallah, M.A.E.; Ali, N.; Law, R.J.; Herzke, D.; de Wit, C.A. Novel brominated flame retardants: A review of their analysis, environmental fate and behavior. *Environ. Int.* **2011**, *37*, 532. [CrossRef]

20. de Wit, C.A.; Herzke, D.; Vorkamp, K. Brominated flame retardants in the Arctic environment—Trends and new candidates. *Sci. Total Environ.* **2010**, *15*, 2885. [CrossRef]

21. Levchik, S.V. A Review of Recent Progress in Phosphorus-based Flame Retardants. *J. Fire Sci.* **2006**, *24*, 345. [CrossRef]

22. Levchik, S.V.; Weil, E.D. Developments in phosphorus flame retardants. In *Advances in Fire Retardant Materials*; Horrocks, A.R., Price, D., Eds.; Woodhead Publishing Ltd.: Cambridge, UK, 2007; pp. 41–66, ISBN 9781845692629.

23. Schartel, B. Phosphorus-based Flame Retardancy Mechanisms—Old Hat or a Starting Point for Future Development? *Materials* **2010**, *3*, 4710. [CrossRef] [PubMed]

24. Velencoso, M.M.; Battig, A.; Markwart, J.C.; Schartel, B.; Wurm, F.R. Molecular Firefighting—How Modern Phosphorus Chemistry Can Help Solve the Challenge of Flame Retardancy. *Angew. Chem. Int. Ed.* **2018**, *57*, 10450–10467. [CrossRef] [PubMed]

25. Kandola, B.K. *Flame retardant characteristics of natural fibre composites In Natural Polymers Volume 1: Composites*; John, M.J., Thomas, S., Eds.; The Royal Society Chemistry: Cambridge, UK, 2012; pp. 108–110, ISBN 9781849734028.

26. Schartel, B.; Bartholmai, M.; Knoll, U. Some comments on the use of cone calorimeter data. *Polym. Degrad. Stab.* **2005**, *88*, 540. [CrossRef]

27. Costes, L.; Laoutid, F.; Brohez, S.; Dubois, P. Bio-based flame retardants: When nature meets fire protection. *Mater. Sci. Eng. R* **2017**, *117*, 1. [CrossRef]

28. Reddy, M.M.; Vivekanandhan, S.; Misra, M.; Bhatia, S.K.; Mohanty, A.K. Biobased plastics and bionanocomposites: Current status and future opportunities. *Prog. Polym. Sci.* **2013**, *38*, 1653. [CrossRef]

29. Xu, Z.H.; Kong, Z.N. Mechanical and thermal properties of short-coirfiber-reinforced natural rubber/polyethylene composites. *Mech. Compos. Mater.* **2014**, *50*, 353. [CrossRef]

30. Pickering, K.L.; Aruan Efendy, M.G.; Le, T.M. A review of recent developments in natural fibre composites and their mechanical performance. *Compos. Part A Appl. Sci. Manuf.* **2016**, *83*, 98. [CrossRef]

31. Kiliaris, P.; Papaspyrides, C.D. Polymer/layered silicate (clay) nanocomposites: An overview of flame retardancy. *Prog. Polym. Sci.* **2010**, *35*, 902. [CrossRef]

32. Kakroodi, A.R.; Kazemi, Y.; Rodrigue, D. Mechanical, rheological, morphological and water absorption properties of maleated polyethylene/hemp composites: Effect of ground tire rubber addition. *Compos. Part B Eng.* **2013**, *51*, 337. [CrossRef]

33. Le Moigne, N.; van den Oever, M.; Budtova, T. Dynamic and capillary shear rheology of natural fiber-reinforced composites. *Polym. Eng. Sci.* **2013**, *53*, 2582. [CrossRef]

34. Ho, M.; Wang, H.; Lee, J.; Lau, K.; Leng, J.; Hui, D. Critical factors on manufacturing processes of natural fibre composites. *Compos. Part B Eng.* **2012**, *43*, 3549. [CrossRef]

35. Borchani, K.E.; Carrot, C.; Jaziri, M. Biocomposites of Alfa fibers dispersed in the Mater-Bi® type bioplastic: Morphology, mechanical and thermal properties. *Compos. Part A Appl. Sci. Manuf.* **2015**, *78*, 371. [CrossRef]

36. Schartel, B.; Wilkie, C.A.; Camino, G. Recommendations on the scientific approach to polymer flame retardancy: Part 2—Concepts. *J. Fire Sci.* **2017**, *35*, 3. [CrossRef]

37. Horrocks, A.R.; Smart, G.; Nazaré, S.; Kandola, B.; Prince, D. Quantification of zinc hydroxystannate** and stannate** synergies in halogen-containing flame-retardant polymeric formulations. *J. Fire Sci.* **2010**, *28*, 217. [CrossRef]

38. Schartel, B.; Hull, T.R. Development of fire-retarded materials—Interpretation of cone calorimeter data. *Fire Mater.* **2007**, *31*, 327. [CrossRef]

 *materials*

Article

Influence of Ammonium Polyphosphate/Lignin Ratio on Thermal and Fire Behavior of Biobased Thermoplastic: The Case of Polyamide 11

Aurélie Cayla [1], François Rault [1], Stéphane Giraud [1,*], Fabien Salaün [1], Rodolphe Sonnier [2] and Loïc Dumazert [2]

[1] ENSAIT, GEMTEX—Laboratoire de Génie et Matériaux Textiles, F-59000 Lille, France; aurelie.cayla@ensait.fr (A.C.); francois.rault@ensait.fr (F.R.); fabien.salaun@ensait.fr (F.S.)

[2] IMT Mines d'Alès, Centre des Matériaux des Mines d'Alès–Pôle Matériaux Polymères Avancés, 30100 Alès, France; rodolphe.sonnier@mines-ales.fr (R.S.); loic.dumazert@mines-ales.fr (L.D.)

* Correspondence: stephane.giraud@ensait.fr; Tel.: +33-320-256-464

Received: 17 March 2019; Accepted: 7 April 2019; Published: 8 April 2019

Abstract: Flame retardancy of polymers is a recurring obligation for many applications. The development trend of biobased materials is no exception to this rule, and solutions of flame retardants from agro-resources give an advantage. Lignin is produced as a waste by-product from some industries, and can be used in the intumescent formation development as a source of carbon combined with an acid source. In this study, the flame retardancy of polyamide 11 (PA) is carried out by extrusion with a kraft lignin (KL) and ammonium polyphosphate (AP). The study of the optimal ratio between the KL and the AP makes it possible to optimize the fire properties as well as to reduce the cost and facilitates the implementation of the blend by a melting process. The properties of thermal decomposition and the fire reaction have been studied by thermogravimetric analyzes, pyrolysis combustion flow calorimetry (PCFC) and vertical flame spread tests (UL94). KL permits a charring effect delaying thermal degradation and decreases by 66% the peak of heat release rate in comparison with raw PA. The fire reaction of the ternary blends is improved even if KL-AP association does not have a synergy effect. The 25/75 and 33/67 KL/AP ratios in PA give an intumescence behavior under flame exposure.

Keywords: lignin; polyamide 11; ammonium polyphosphate; thermal decomposition; fire reaction

---

1. Introduction

Until now the plastic industry was governed by the use of petroleum resources. However, an evolution of the sector is underway. In the last few years, a transition has taken place between fully fossil-based polymers and fully-biobased polymers being replaced by partially biobased ones. Biobased polymers can be considered as macromolecular materials coming from biological resources and transformed by humans to be used in various activities in the form of massive sheets, films, or fibers. Biobased polyethylene (PE), polylactic acid (PLA), or even polyamide 11 (PA) derived from castor oil are some examples of novel thermoplastic polymers able to compete with conventional polymers. Nevertheless, in order to ensure their development and their use on the market, only emphasizing the environmental benefits for those polymers compared to those derived from petroleum resources is not enough. Indeed, these polymers must satisfy the same prerequisites as their predecessors. They have to reach the same criteria related to health, legislation, and the economy, as well as various performance criteria. The reduction of fire risks is one of the properties required and sometimes mandatory to access sectors such as transport and buildings. One of the challenges is then to guarantee the security of the proposed solutions while preserving as much as possible the bio-sourced aspect. Thus in recent years some of the research related to the flame retardancy of polymers has focused on the use of "green" flame retardant additives [1].

Different strategies exist and/or co-exist in order to fireproof petroleum-based polymers and can be transposed to the biobased ones. Two modes of actions, i.e., in the gas phase or condensed phase, may operate independently or together. Their mechanisms require radical inhibition or dilution of the combustion gas in the first mode, and endothermic degradation or dilution of the polymer, or even creation of a physical and/or thermal protective barrier, in the second mode. The latter approach refers mainly to the concept of intumescence and consequently to intumescent systems. An acidic source, a blowing or swelling agent, and a carbonizing agent are traditionally involved in the development of such systems that lead to the formation of an intumescent char (expanded carbonaceous barrier) upon heating. One of the most studied formulations is based on phosphorus-based additives able to release phosphoric acid, polyols as a carbon source and melamine as a blowing agent. These formulations can sometimes be limited to the use of only two components. Indeed, some components such as ammonium polyphosphate (AP) can act as an acid source and a swelling agent. Many studies have therefore been interested in developing fully (or at least partially) biobased intumescent formulations. In the case of biobased polymers, the use of natural additive is, even more so, an obvious solution to develop fire retardant formulation. It allows preservation of the original philosophy of sustainable development and the production of eco-friendly new high value-added polymers. Thus, for the first time, scientists have especially tried to substitute carbon sources traditionally used for biobased polymers. Formulations combining AP and alternative natural-based carbon sources such as starch [2–4], β-cyclodextrin [5–7], and chitosan [8] have been tested with one of the most investigated biobased polymers, i.e., PLA. Among the other natural-based compounds such as DNA, proteins, other saccharide-based compounds (sorbitol, xylitol, etc.), or derivatives from them (ex: isosorbide), biobased aromatic products, and in particular lignin, are of great interest to develop intumescent formulations. It is mentioned that its chemical structure (content of hydroxyl groups and aromatic structure) promotes the formation of a high char yield after decomposition and that its fire-retardant effect can be improved by associating other flame-retardant components [9]. Furthermore, lignin is an abundant co-product of the pulp and paper industry. Therefore, it can be considered as a waste of them and consequently provide a low-cost, eco-friendly alternative to petroleum-based carbon sources in intumescent systems for bioplastics. Its impact on the modification of the fire behavior of the polybutylene succinate [10,11], polylactic acid [2,12–17], or polyamide 11 [18–20] has been evaluated.

The flame retardancy of a biobased polymer such as PLA is relatively well documented in the literature. Different scientists have worked on this issue by considering several strategies including the use of intumescent systems that may be or not bio-sourced. In contrast, the fire-retardant enhancement of other biobased thermoplastic by using the concept of intumescence, in particular in polyamide 11, has up to now more rarely been investigated. Levchick et al. [21] studied the fire behavior of polyamide 11 with AP. One of the main conclusions is that the yield of residue formed seems not to be significantly influenced by AP. To the best of our knowledge, our research group is the first to consider the development of an intumescent flame retardant system for polyamide 11 with a double target, i.e., i) push the "green" concept as far as possible, and ii) keep the additives content compatible with further textile fiber forming process. The purpose of this article is to study the combustion behavior of polyamide 11 containing intumescent system composed of lignin and AP. Blends of PA with different kraft lignin (KL)/AP ratio were prepared by melt extrusion. UL-94 and pyrolysis combustion flow calorimetry (PCFC) experiments evaluated the flame-retardant effectiveness of such a solution. Furthermore, the potential synergy between lignin and AP as well as their interactions with polyamide 11 were investigated. Even if the thermal behavior of the KL/AP powder mixtures is different with the KL/AP ratio, the PA blends with the same mixtures present a quite similar thermal and fire behavior without synergy compared to PA blends with only AP or KL.

2. Materials and Methods

2.1. Materials and Processing

A biobased Polyamide 11 (PA), Rilsan®BMNO-TLD; Mn = 17,000 g/mol, melt flow index (MFI) =14–20 g/10 min at 235 °C, supplied from Arkema (Colombes, France), was chosen as the polymer matrix. The lignin used as charring agent is a kraft lignin (KL) purchased from Sigma Aldrich (Darmstadt, Germany). Ammonium polyphosphate (AP) with the reference Exolit AP 422 was supplied by Clariant (Muttenz, Switzerland). The both additives consist of a thin powder with an averaged particle size of 39 µm and 15 µm respectively. All the materials were dried at 80 °C for 24 h before any use.

2.2. Blends Preparation

The compositions of all the samples are listed in Table 1. In order to understand the influence of coupled additives on PA, the thermal decomposition of simple powder mixtures of both additives was assessed. Blends with PA were prepared with a suitable quantity of polymer and additives (lignin and/or ammonium polyphosphate). In order to obtain homogeneous pellets, each premix was extruded in a co-rotating, intermeshing twin screw extruder Thermo Haake (diameter of screw = 16 mm, L/D ratio = 25) from Thermo Fisher Scientific (Waltham, MA, USA). The rotating speed was maintained at 100 rpm. The five heating zones of the extruder were set at temperatures 170, 190, 200, 220, and 220 °C respectively to keep the highest fluidity without damaging the components. The different formulations extruded (with or without fillers) were then cooled under ambient air and pelletized. The pellets obtained for each formulation were transformed in plates of $100 \times 13 \times 3$ mm^3 (for UL94 tests) using a heat press from Dolouets (Soustons, France) at 220 °C under 50 bars.

Table 1. Blend formulations: kraft lignin (KL)-ammonium polyphosphate (AP) powder mix and polyamide 11 (PA) composites.

Sample	PA (wt. %)	KL (wt. %)	AP (wt. %)
KL$_{25}$-AP$_{75}$	-	25	75
KL$_{33}$-AP$_{67}$	-	33	67
KL$_{50}$-AP$_{50}$	-	50	50
PA$_{100}$	100	-	-
PA$_{80}$-KL$_{20}$	80	20	-
PA$_{80}$-AP$_{20}$	80	-	20
PA$_{80}$-KL$_{05}$-AP$_{15}$	80	5	15
PA$_{80}$-KL$_{07}$-AP$_{13}$	80	7	13
PA$_{80}$-KL$_{10}$-AP$_{10}$	80	10	10

2.3. Thermal Decomposition

Thermogravimetric analyses were performed using a TA instruments thermal analyzer model number 2050 (New Castle, DE, USA) to study the thermal stability of blends. The sample of 6 ± 1 mg was placed in an open platinum crucible, and an empty platinum crucible was used as the reference. Dynamic runs were carried out from room temperature to 600 or 800 °C, at a heating rate of 10 °C/min in a 50 mL/min flow of nitrogen. Thermogravimetric curves (TG) curves were recorded from experiments, and Derivative thermogravimetric (DTG) curves were obtained from TA universal data analysis software for all the samples. The decomposition parameters, such as the temperature at 5% and 50% weight loss ($T_{5\%}$ and $T_{50\%}$), and residue at 600 or 800 °C were obtained from the TG curve. Furthermore, the maximum mass loss rate (R_{max}) and the corresponding temperature (T_{max}) for each main degradation step were obtained from DTG curves.

The residual mass difference curves were plotted in order to determine the increase or decrease in the thermal stability of the blends with the subsequent interaction between components. The residual mass difference curves were computed for the powder additive mixtures and composites samples, and correspond to the residual mass difference between the experimental and theoretical TG curves Equation (1).

$$\Delta(M(T)) = M_{\exp}(T) - M_{theo}(T),\qquad(1)$$

where, $\Delta(M(T))$ is a residual mass difference, $M_{\exp}(T)$ is the experimental residual mass of blends (variation with temperature T), $M_{theo}(T)$ is the theoretical residual mass of the same composition computed by a linear combination between the experimental masses of each components. The residual mass difference of the blend at the temperature T permits to give an indication on thermal stability of the blend regarding the whole thermal history until the temperature T.

2.4. Fire Reaction

The flammability was evaluated on sample bars ($125 \times 12.5 \times 3$ mm^3) by vertical flame spread tests according to IEC 60695-11-10 [22], also known as UL 94 burning flame test and used for plastic materials. This test is aimed at assessing the material capability to extinguish a flame. Materials were classified from their burning rate, time to flame out and dripping during burning.

We also employed pyrolysis combustion flow calorimetry (PCFC) (Fire Testing Technology, East Grinstead, UK). This technique was described by Lyon [23]. Sample amount of some milligrams is pyrolyzed with a heating ramp (1 K/s) up to 750 °C under nitrogen flow. The gases released during the pyrolysis are removed in an oven at 900 °C in the presence of an 80/20 N_2/O_2 mixture. In these conditions, the total combustion of these gases takes place. According to Huggett's relation (1 kg of consumed O_2 corresponds to 13.1 MJ of released energy), the measurement of oxygen consumption by PCFC permits to calculate the heat release rate. Three tests were carried out on each formulation, and the results averaged. According to this analytical method, usual parameters were evaluated, i.e., the peak of heat release rate (pHRR) and its temperature, total heat release (THR), char residue and heat of complete combustion (Δh) which is calculated by the division of the THR on mass loss fraction.

3. Results and Discussion

3.1. Thermal Degradation

Lignin is composed of many components having different decomposition pathways leading its thermal degradation following a complex process with consecutive reactions [24]. Thus, lignin decomposes slower over a broad temperature range from 200 to 500 °C, and the DTG curve shows a wide and flat peak with a gentle slope line (Figure 1), due to the presence of various oxygen functional groups in its structure. Each group has different thermal stability, and therefore their scission occurs at different temperatures. During the thermal decomposition, the cleavages of the lignin functional groups lead to the formation of low molecular weight products until the chemical rearrangement of the backbone at a higher temperature with the formation of a significant char and the release of volatile products. Up to 200 °C, lignin has an initial weight loss equal to around 5%, mainly due to moisture release and volatile impurities. Above this temperature, the compound undergoes a two steps degradation process under an inert atmosphere. From 200 to 500 °C, heated up by 10 °C·min^{-1}, lignin decomposes very slowly, since the max rate of degradation at 355 °C is only about 0.2433 %·°C^{-1}, losing only 48 wt. % of its initial mass below 500 °C (Table 2). During this main degradation step, the cleavage of the aryl-ether linkages results in the formation of aromatic hydrocarbons, phenolic, hydroxyphenolics, and guaiacyl/syringyl-type compounds according to Rodrigues et al. [25]. Thereafter, at the end of the pyrolysis phenols groups are transformed into pyrocatechols [26]. The degradation rate slightly decreases to 0.03 %·°C^{-1} during the second step, with the formation of a residue of about 43 wt. % at 800 °C due to the formation of polycyclic aromatic hydrocarbons.

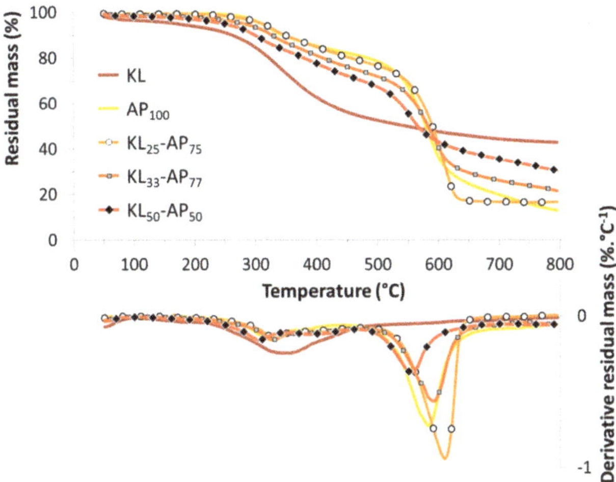

Figure 1. TG and DTG curves of KL, AP, and their blends.

Table 2. TGA results of lignin, ammonium polyphosphate and their respective blends under nitrogen atmosphere.

Samples	$T_{5\%}$ (°C)	$T_{50\%}$ (°C)	T_{max} (°C) & R_{max} (%·°C^{-1})			Residue (%)[1]
			Step I	Step II	Step III	800 °C
KL	185	551	355 0.2433	-	-	43.0
AP	311	586	327 0.1345	583 0.7235	-	13.0
KL$_{50}$-AP$_{50}$	260	570	305 0.1571	386 0.1210	557 0.3696	33.0 (28.0)
KL$_{33}$-AP$_{67}$	280	578	328 0.1557	593 0.5613	-	23.7 (23.0)
KL$_{25}$-AP$_{75}$	314	590	333 0.1487	611 0.9401	-	16.5 (20.5)

[1] in brackets calculated values based on additive behavior.

The thermal decomposition of ammonium polyphosphate occurs in two-steps mechanism in nitrogen conditions [27]. The first degradation step from 200 to 450 °C, with 18% of weight loss, involves the water and ammonia releases as the gaseous products. Maximum degradation peak was observed at approximately 327 °C. The water elimination further induces the formation of phosphoric acid coupled to a cross-linking mechanism to lead the formation of polyphosphoric acid. The second degradation step from 450 to 700 °C is related to the dehydration and the fragmentation of polyphosphoric acid to form phosphorus oxides. In this temperature range, this material undergoes a sharp weight loss (69 wt. %) with the formation of a stable residue (13 wt. %) upon heating up to 800 °C.

The presence of ammonium polyphosphate changed the thermal behavior of lignin (Figure 1 and Table 2). Regardless of the ratio, $T_{5\%}$ and $T_{50\%}$ were found to be increased. Compared with pure KL, the higher initial degradation temperature is due to the higher thermal stability of AP, while the much higher $T_{50\%}$ (improvement of thermal stability) is mainly attributed to a much higher char residue. Furthermore, the thermal degradation occurs in two consecutive steps in the same temperature ranges than the AP one for the samples KL$_{25}$-AP$_{75}$ and KL$_{33}$-AP$_{67}$. According to the DTG curves, the maximal rate of weight loss for two-step degradation is at 328 and 593 °C, 333 and 611 °C for KL_{33}-AP_{67} and KL_{25}-AP_{75}, respectively. The decomposition shift to a higher temperature may be attributed to the increasing content of AP. It can also be noticed the presence of an intermediate step for the sample labeled KL_{50}-AP_{50} between 350 and 470°C. The first step degradation of KL-AP blends occurs at a lower temperature compared to KL, whereas the second step is shifted towards higher temperature compared to AP, except for the sample labeled KL_{50}-AP_{50}. Above 200°C, the mass loss rates decrease in comparison to that for KL leading to stabilization until the second degradation step. Thus, R_{max} is indeed reduced from 0.24 to 0.13 or 0.15 wt. %·°C^{-1} under pyrolytic conditions. It appears that the adding of AP induces a slight thermal destabilization in the first step decomposition because phosphoric acid molecules catalyzed dehydration of lignin. For the second or third (sample for KL_{50}-AP_{50}) degradation step, R_{max} values are found to be lower than for AP when AP content is less than 75% in weight, and higher for this sample. At high temperature, the residue amount increases from 16.5 to 33 wt. % as the AP content decreases.

Figure 2 allows to point out the interactions between the decomposition products of KL and AP during thermal degradation of the KL-AP mixtures. The KL_{25}-AP_{75} and KL_{33}-AP_{67} blends present until 600 °C a thermal stabilization phase where the residual mass of the blend is higher than the addition of the residual mass of KL and AP. This positive interaction between the decomposition products of KL and AP has two local maxima at 390 °C and 570 °C. After 600 °C, both blends are thermally destabilized with minima at 630 °C. The positive and negative interactions are more amplified for KL_{25}-AP_{75} (maximum +7.5%/minimum −13.7%) than for KL_{33}-AP_{67} (maximum +3.3%/minimum −2.2%). The KL_{50}-AP_{50} blend also has a maximum positive interaction +3.5% at 390°C, but between 530 and 600 °C, the blend presents a thermal destabilization with −4% minimum. Above 600 °C, the decomposition products interact again positively with +3.5% maximum at 710 °C. Thus, smaller AP contents lead to a higher charring effect with better thermal stabilization.

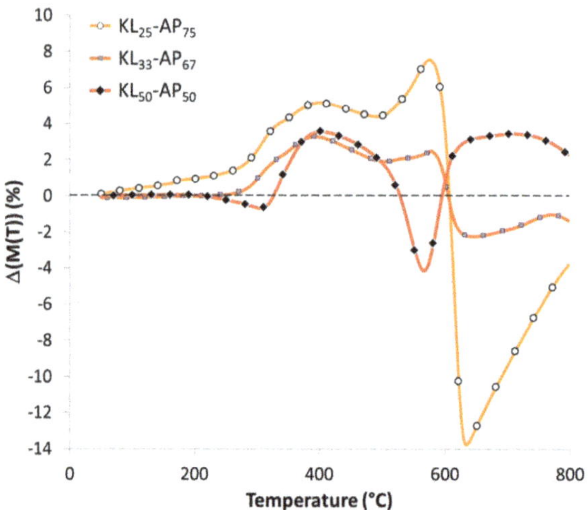

Figure 2. Curves of residual mass loss difference for KL-AP powder blends.

TG and DTG curves of PA and PA/lignin and/or ammonium polyphosphate blends under N_2 atmosphere are shown in Figure 3. It can be observed that the neat PA shows a two-step thermal degradation in the range of 350–500°C with no char residue left at 600 °C. The primary step is located around 430 °C, whereas the second and minor step occurs over 450 °C due to the decomposition of cross-linked structures obtained during heating, as a shoulder in the DTG curve suggests [21]. For PA with 20% of AP, the onset temperature and the temperature at the maximum rate of weight loss (Table 3) are shifted toward low temperature by respectively 30 and 40 °C compared to raw PA, due to the early decomposition of AP. The polyphosphoric acid released from the ammonium polyphosphate decomposition reacted with the amide bond to form intermediate phosphate ester bonds, leading to a decrease at $T_{50\%}$ of PA by 40 °C. A substantial thermal destabilization for the blend PA_{80}-AP_{20} is also observed on the differential mass curve (Figure 4) between 350 and 500 °C. The phosphate ester bonds decomposed at a higher temperature to promote the formation of char, it can be observed that the obtained residue at 600 °C is slightly higher than the theoretical values (Table 3).

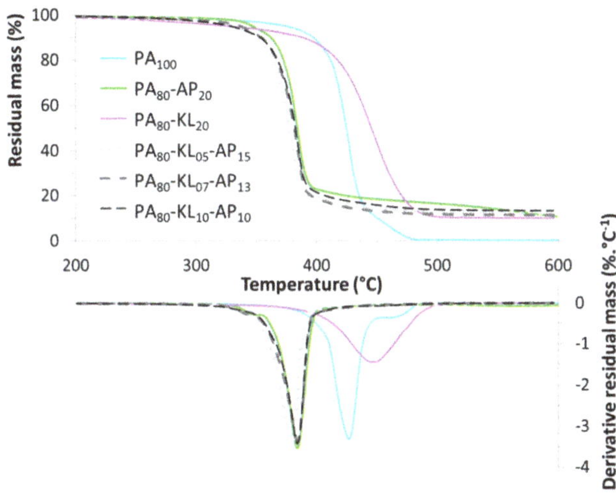

Figure 3. TG and DTG curves of PA and PA blends with AP and KL under nitrogen atmosphere.

Table 3. TGA results of PA and blends with lignin and/or ammonium polyphosphate under nitrogen atmosphere.

Samples	$T_{5\%}$ (°C)	$T_{50\%}$ (°C)	T_{max} (°C) & R_{max} (%·°C^{-1})			Residue (%)[1]
			Step I	Step II	Step III	600 °C
PA_{100}	375	424	426 3.320	463 0.361	-	0.2
PA_{80}-AP_{20}	347	383	384 3.485	-	-	10.7 (7.3)
PA_{80}-KL_{20}	325	445	445 1.443	-	-	10.0 (9.5)
PA_{80}-KL_{05}-AP_{15}	335	377	379 3.888	-	-	12.3 (8.4)
PA_{80}-KL_{07}-AP_{13}	336	381	384 3.381	-	-	11.3 (8.1)
PA_{80}-KL_{10}-AP_{10}	329	382	384 3.431	-	-	13.3 (8.6)

[1] in brackets calculated values based on additive behavior.

The presence of lignin at 20% in PA is responsible for the decrease of $T_{5\%}$ by 50 °C, and an increase of $T_{50\%}$ and T_{max} about 20 °C. Furthermore, the thermal degradation occurs in a single, broader stage. The decrease of $T_{5\%}$ may be attributed to a rapid mass loss rate of KL at a lower temperature region. The difference between TG curves of raw and filled polymer (Figure 4) shows a noticeable thermal stabilization between 400 and 500 °C. So KL permits a charring effect delaying thermal degradation of the blend in comparison with raw PA. However, above 500 °C the residual mass difference of the blend is 0 and the residue amounts correspond to the KL content after its degradation. Thus, it can be concluded that there is no interaction between decomposition products of PA and KL.

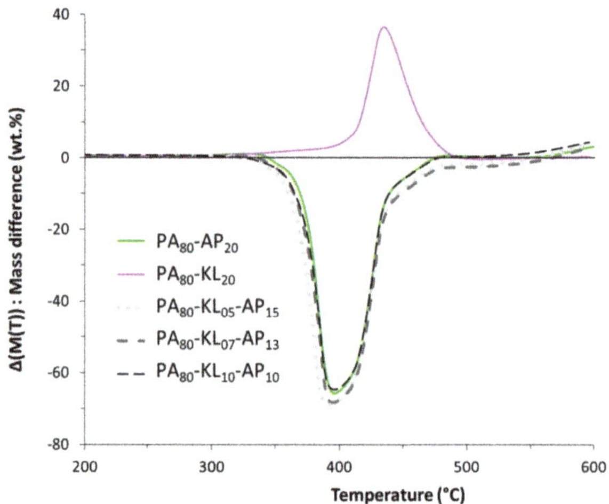

Figure 4. Curves of residual mass loss difference for PA blends with AP and KL.

The mass loss behavior of PA_{80}-KL_{05}-AP_{15}, PA_{80}-KL_{07}-AP_{13}, and PA_{80}-KL_{10}-AP_{10} samples (Figure 3) are almost similar, and their DTG maxima are observed at 379 and 384 °C, respectively. The influence of AP in the thermal degradation of the ternary blends is predominant since their TG curve is closed to the TG curve of PA_{80}-AP_{20}. The onset temperature, $T_{5\%}$, of the ternary blends are found to be decreased compared to raw PA and also PA-AP samples, by 40 to 50 °C and 11 to 18 °C, respectively. Their decomposition temperatures (T_{max}) are decreased compared to that of raw PA, whereas the maximum rates of degradation are found to be slightly increased for the samples PA_{80}-KL_{07}-AP_{13} and PA_{80}-KL_{10}-AP_{10}, and more significantly for the PA_{80}-KL_{05}-AP_{15} one. Therefore, the presence of AP has a catalyzing effect on the degradation of PA, which is stronger in presence of KL in the initial step of degradation. Furthermore, the TG curve of KL shows that 7% of KL decomposes at 220 °C, i.e., at processing temperature to prepare PA-KL-AP blends. Thus, the decrease of $T_{5\%}$ of these blends by 30 to 40 °C is due to early decomposition of KL and AP. As for mass loss difference of PA_{80}-AP_{20} sample, the mass loss difference of the three ternary blends (Figure 4) presents a critical thermal destabilization period from 350 and 500 °C. It can also be noticed that the thermal degradation of the three ternary blends seems to be irrelevant of the AP to KL weight ratio used in this study, since the change trends of TG curves between all samples are similar, even if there are still some differences. The initial decomposition temperature of PA_{80}-KL_{10}-AP_{10} is lower than the two other samples ones. Besides, for a sufficient AP content, when samples begin to degrade, the charring aromatic radicals coming from KL reduce the polymer degradation rate increasing the composite thermal degradation temperature by 5 °C. On the other hand, the residue left at 600 °C for PA_{80}-KL_{05}-AP_{15}, PA_{80}-KL_{07}-AP_{13}, and PA_{80}-KL_{10}-AP_{10} are about 12.3, 11.3, and 13.3, respectively, suggesting higher charring than PA-KL and PA-AP samples. Therefore, the increase in the amount of residue may be owing to the formation of more stable carbonaceous char. Indeed, the char yield for the three ternary blends at 600 °C (Table 3) is higher than the sum of the individual contributions of each component. Thus, more effective carbonizing and cross-linking reactions take place during the PA degradation on the addition of AP with KL in comparison to KL or AP solely. From the above results, It seems that the blend with the 50:50 weight ratio of KL and AP is better than the two other ones (25:75 and 33:67) in improving the charring of the PA composite.

3.2. Flammability Behavior

The results of the UL94 tests for PA and the blends are summarized in Table 4, and the typical pictures of the specimens left after the tests are shown in Figure 5.

Table 4. UL94 vertical flame spread test data for PA and its blends.

Samples	1st Flame t_1 (s)	2nd Flame t_2 (s)	Combustion Time (t_1+t_2)	Cotton Ignition	Rating	Mass Loss (%)
PA_{100}	1.6 ± 0.4	2.4 ± 0.6	4.0 ± 1.0	Yes	V2	33.7 ± 10.9
PA_{80}-AP_{20}	8.9 ± 5.9	4.8 ± 1.6	13.7 ± 5.7	Yes	V2	34.7 ± 4.9
PA_{80}-KL_{20}	1.1 ± 0.1	14.4 ± 11.3	15.5 ± 11.5	Yes	V2	64.1 ± 3.3
PA_{80}-KL_{05}-AP_{15}	1.9 ± 0.9	2.7 ± 1.4	4.6 ± 1.9	Yes	V2	21.4 ± 3.3
PA_{80}-KL_{07}-AP_{13}	3.0 ± 2.2	7.0 ± 1.9	10.0 ± 1.2	Yes	V2	19.1 ± 4.5
PA_{80}-KL_{10}-AP_{10}	6.6 ± 0.7	22.8 ± 5.9	29.4 ± 6.5	Yes	V2	81.5 ± 16.6

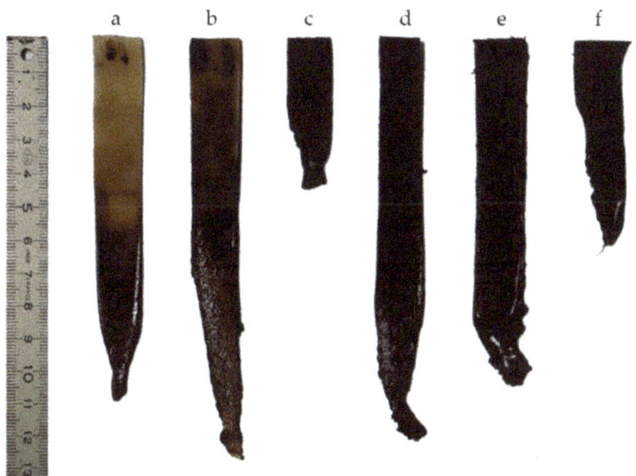

Figure 5. Pictures of PA blends specimen left after UL 94 vertical flame test. (**a**) PA; (**b**) PA_{80}-AP_{20}; (**c**) PA_{80}-KL_{20}; (**d**) PA_{80}-KL_{05}-AP_{15}; (**e**) PA_{80}-KL_{07}-AP13; (**f**) PA_{80}-KL_{10}-AP_{10}.

Dripping and cotton ignition were observed in all the blends. Just after the flame exposure, PA presents a significant dripping of burning materials. As already described in literature for some thermoplastic polymers like Polyamide 6 [28], the dripping is so consequent that the flame spread is limited and so the remaining preserved material is noticeable. The presence of AP or/and KL does not give significant improvement in the burning behavior of PA. With 20% of AP, we can observe during flame exposition a low charring effect with some crackling due to gas action of AP (ammonia and water release). Contrary to raw PA, the combustion is maintained during 10 s more in total after removing of the flame. The flame behavior of thermoplastic polymers in the vertical direction is complex, not only the mechanisms of thermal degradation have a determining role, but the viscoelastic properties in the molten state are also crucial [29]. The viscosity of molten PA_{80}-AP_{20} blend is higher than for PA (Table S1). This difference of viscosity could be not favorable to PA_{80}-AP_{20} blend from a flame spread point of view. The combustion stopped finally by dripping of burning materials. The combustion time for PA_{80}-KL_{20} after the first flame exposure is short due to a rapid dripping of burning material, but after the 2nd exposure, the dripping is slowed down, and a small flame spread during almost 15 s destroyed two-thirds of the samples. The best results are obtained with PA_{80}-KL_{05}-AP_{15} and PA_{80}-KL_{07}-AP_{13} which show a significant flame behavior improvement. A charring effect with some intumescence is observable during flame exposure, and the material is self-extinguishing after the first exposure. During the second exposure, some burning material drips but the samples kept finally about 80%

of the initial mass. In the case of PA_{80}-KL_{10}-AP_{10}, the charring effect is lower than in the case of the two other ternary blends, and no intumescence is observed, allowing the flame to spread on almost the entire sample.

3.3. Combustion Behavior

The PCFC experiments provide access to the heat release rate from the complete combustion of fuel released during the anaerobic pyrolysis of the material. Table 5 presents for each PA blend the peak of heat release rate with its temperature, the total heat release, the char residue, and the heat of complete combustion. The Figure 6 shows the curves of HRR. HRR curves are in good agreement with DTG curves in Figure 3. Indeed, pyrolysis conditions are similar in PCFC and TGA in nitrogen.

Table 5. Pyrolysis combustion flow calorimetry (PCFC) data for PA and its blends. pHRR: peak of heat release rate; THR: total heat release; Δh: heat of complete combustion.

Samples	pHRR (W/g)	pHRR Reduction (%)	pHRR Temperature (°C)	THR (KJ/g)	THR Reduction (%)	Residue (%)	Δh (KJ/g)
PA_{100}	1293 ± 39	-	413 ± 2	32.4 ± 0.1	-	3.2 ± 0.3	33.6 ± 0.4
PA_{80}-AP_{20}	1108 ± 92	14	382 ± 3	29.3 ± 0.4	10	11.4 ± 0.0	33.0 ± 0.4
PA_{80}-KL_{20}	442 ± 5	66	451 ± 2	28.3 ± 0.1	13	11.0 ± 0.2	31.7 ± 0.1
PA_{80}-KL_{05}-AP_{15}	993 ± 30	23	380 ± 1	27.9 ± 0.2	14	14.4 ± 0.4	32.6 ± 0.4
PA_{80}-KL_{07}-AP_{13}	908 ± 47	30	379 ± 0	26.8 ± 0.5	17	13.9 ± 0.1	31.1 ± 0.6
PA_{80}-KL_{10}-AP_{10}	924 ± 19	29	383 ± 2	26.9 ± 0.9	17	12.1 ± 0.3	30.5 ± 0.4

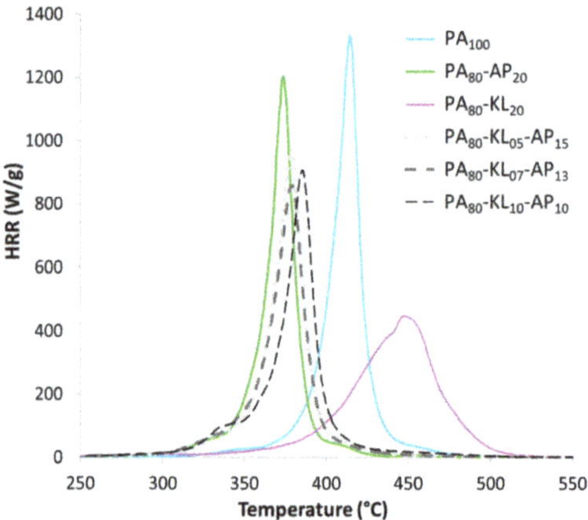

Figure 6. Curves of heat release rate (HRR) from PCFC experiments for PA blends with AP and KL.

Among the PA blends with KL and/or AP, the PA_{80}-KL_{20} blend presents the best behavior combustion since the pHRR is decreased by 66% and shifted by around +40 °C in comparison of raw PA. The PCFC results for PA_{80}-KL_{20} blend are correlated to the fact that the thermal degradation with 20% of KL is delayed, and slowed down compared to raw PA. However, the residue content for the PA_{80}-KL_{20} blend remains quite low, and so the total heat released of PA_{80}-KL_{20} is only slightly decreased (13%) compared to the THR of raw PA. The decrease of the Δh for PA_{80}-KL_{20} blend (31.7 kJ/g) is limited (33.6 kJ/g for raw PA) and can be assigned to the replacement of a fraction of PA by KL having a low heat of combustion (around 10 kJ/g) [30]. Given PCFC results, the PA_{80}-AP_{20} shows the lowest improvement of flame retardant effect. The HRR peak, the THR and the Δh of PA_{80}-AP_{20} are closed to the raw PA values. Moreover, since PA_{80}-AP_{20} has lower thermal stability than PA, the pHRR for PA_{80}-AP_{20} is 30 °C below. The PCFC results for the three ternary blends are slightly better than for PA_{80}-AP_{20} blend. Once again, residue content is slightly enhanced compared to binary blends. Therefore THR is reduced. Nevertheless, the destabilization due to AP leads to a pHRR at low temperature (around 30 °C below than that of PA, and the pHRR remains very high even if it is slightly reduced compared to raw PA and PA_{80}-AP_{20}.

4. Conclusions

The thermogravimetric analyses of the kraft lignin/ammonium polyphosphate powder mixtures, respectively charring agent and acidic source for potential flame retardant intumescent formulation point out noticeable interactions between both components whatever the ratio of the mixture. However, the KL-AP ratio influences these interactions. The curves of residual mass loss difference for KL-AP powder blends reveal a positive interaction between around 300 and 500 °C. However, unlike the KL_{25}-AP_{75} and KL_{33}-AP_{67} mixtures, only the KL_{50}-AP_{50} mixture which has the highest residue at 800 °C keeps a positive interaction above 600 °C. The influence of KL-AP ratio on the thermal decomposition of the PA-KL-AP blends is low. The thermal degradations of the different ternary blends are similar and close to the decomposition of the PA-AP blend. AP finally dominates the degradation of the ternary blends. On the other hand, the presence of 20% of KL alone in PA has a positive effect on the thermal degradation with a delay in the main PA degradation step. The analyze of the fire behavior for the different PA blends by PCFC indicates that the fire retardant capacity of the ternary blends is intermediate considering the peak of HRR between these of the PA-AP and PA-KL blends. If the PA-KL blend shows the smallest peak of HRR, it is probably due mainly to the low combustion heat of the lignin. The KL-PA blend also presents the worst result with UL94 flammability test and, contrary to the PCFC test where the fire retardant effect by intumescence are not favored, the ternary blends present the best results even if they still V2 ranking due to creep phenomenon. For this test, the KL-AP ratio seems to have of influence since the formation of an efficiently expanded char is observed only for the PA_{80}-KL_{05}-AP_{15} and PA_{80}-KL_{07}-AP_{13} blends. In the case of PA_{80}-KL_{10}-AP_{10} blend, the flame behavior becomes more similar to the PA_{80}-KL_{20} blend. Even if interactions between AP and KL exist, once both components are dispersed in PA, these interactions develop less. That is why the ternary blends do not present synergy on fire retardant aspect. Nevertheless, the ratio KL-AP affects the PA fire behavior, and given all the results, the PA_{80}-KL_{07}-AP_{13} blend seems to have the best fire retardant.

Supplementary Materials: The following are available online at http://www.mdpi.com/1996-1944/12/7/1146/s1, Table S1: Melting Flow Index for PA and its blends.

Author Contributions: Conceptualization, A.C., F.R., S.G. and F.S.; methodology, A.C., F.R., S.G. and F.S.; software, A.C., F.R., S.G. and F.S.; validation, A.C., F.R., S.G. and F.S.; formal analysis, A.C., F.R., S.G., F.S., R.S. and L.D.; investigation, A.C., F.R., S.G. and F.S.; resources, A.C., F.R., S.G., F.S., R.S. and L.D.; data curation, A.C., F.R., S.G. and F.S.; writing—original draft preparation, A.C., F.R., S.G. and F.S.; writing—review and editing, A.C., F.R., S.G., F.S., and R.S.; visualization, A.C., F.R., S.G. and F.S.; supervision, S.G. and F.S.; project administration, S.G. and F.S.; funding acquisition, A.C., F.R., S.G. and F.S.

Funding: This research received no external funding.

Acknowledgments: The authors gratefully acknowledge Guillaume Lemort (ENSAIT, GEMTEX) for technical support (UL 94 burning flame test).

Conflicts of Interest: The authors declare no conflict of interest.

References

1. Hobbs, C.E. Recent advances in bio-based flame retardant additives for synthetic polymeric materials. *Polymers* **2019**, *11*, 224. [CrossRef]
2. Réti, C.; Casetta, M.; Duquesne, S.; Bourbigot, S.; Delobel, R. Flammability properties of intumescent PLA including starch and lignin. *Polym. Adv. Technol.* **2008**, *19*, 628–635. [CrossRef]
3. Wang, X.; Hu, Y.; Song, L.; Xuan, S.; Xing, W.; Bai, Z.; Lu, H. Flame retardancy and thermal degradation of intumescent flame retardant poly(lactic acid)/starch Biocomposites. *Ind. Eng. Chem. Res.* **2011**, *50*, 713–720. [CrossRef]

4.   Maqsood, M.; Seide, G. Investigation of the flammability and thermal stability of halogen-free intumescent system in biopolymer composites containing biobased carbonization agent and mechanism of their char formation. *Polymers* **2018**, *11*, 48. [CrossRef]

5.   Feng, J.-X.; Su, S.-P.; Zhu, J. An intumescent flame retardant system using β-cyclodextrin as a carbon source in polylactic acid (PLA). *Polym. Adv. Technol.* **2011**, *22*, 1115–1122. [CrossRef]

6.   Wang, X.; Xing, W.; Wang, B.; Wen, P.; Song, L.; Hu, Y.; Zhang, P. Comparative Study on the Effect of Beta-Cyclodextrin and Polypseudorotaxane As Carbon Sources on the Thermal Stability and Flame Retardance of Polylactic Acid. *Ind. Eng. Chem. Res.* **2013**, *52*, 3287–3294. [CrossRef]

7.   Vahabi, H.; Shabanian, M.; Aryanasab, F.; Laoutid, F.; Benali, S.; Saeb, M.R.; Seidi, F.; Kandola, B.K. Three in one: β-cyclodextrin, nanohydroxyapatite, and a nitrogen-rich polymer integrated into a new flame retardant for poly (lactic acid). *Fire Mater.* **2018**, *42*, 593–602. [CrossRef]

8.   Chen, C.; Gu, X.; Jin, X.; Sun, J.; Zhang, S. The effect of chitosan on the flammability and thermal stability of polylactic acid/ammonium polyphosphate biocomposites. *Carbohydr. Polym.* **2017**, *157*, 1586–1593. [CrossRef]

9.   Costes, L.; Laoutid, F.; Brohez, S.; Dubois, P. Bio-based flame retardants: When nature meets fire protection. *Mater. Sci. Eng. R Reports* **2017**, *117*, 1–25. [CrossRef]

10.   Ferry, L.; Dorez, G.; Taguet, A.; Otazaghine, B.; Lopez-Cuesta, J.M. Chemical modification of lignin by phosphorus molecules to improve the fire behavior of polybutylene succinate. *Polym. Degrad. Stab.* **2015**, *113*, 135–143. [CrossRef]

11.   Wu, W.; He, H.; Liu, T.; Wei, R.; Cao, X.; Sun, Q.; Venkatesh, S.; Yuen, R.K.K.; Roy, V.A.L.; Li, R.K.Y. Synergetic enhancement on flame retardancy by melamine phosphate modified lignin in rice husk ash filled P34HB biocomposites. *Compos. Sci. Technol.* **2018**, *168*, 246–254. [CrossRef]

12.   Zhang, R.; Xiao, X.; Tai, Q.; Huang, H.; Hu, Y. Modification of lignin and its application as char agent in intumescent flame-retardant poly(lactic acid). *Polym. Eng. Sci.* **2012**, *52*, 2620–2626. [CrossRef]

13.   Zhang, R.; Xiao, X.; Tai, Q.; Huang, H.; Yang, J.; Hu, Y. Preparation of lignin–silica hybrids and its application in intumescent flame-retardant poly(lactic acid) system. *High Perform. Polym.* **2012**, *24*, 738–746. [CrossRef]

14.   Zhang, R.; Xiao, X.; Tai, Q.; Huang, H.; Yang, J.; Hu, Y. The effect of different organic modified montmorillonites (OMMTs) on the thermal properties and flammability of PLA/MCAPP/lignin systems. *J. Appl. Polym. Sci.* **2013**, *127*, 4967–4973. [CrossRef]

15.   Cayla, A.; Rault, F.; Giraud, S.; Salaün, F.; Fierro, V.; Celzard, A. PLA with intumescent system containing lignin and ammonium polyphosphate for flame retardant textile. *Polymers* **2016**, *8*, 331. [CrossRef]

16.   Costes, L.; Laoutid, F.; Brohez, S.; Delvosalle, C.; Dubois, P. Phytic acid–lignin combination: A simple and efficient route for enhancing thermal and flame retardant properties of polylactide. *Eur. Polym. J.* **2017**, *94*, 270–285. [CrossRef]

17.   Song, Y.; Zong, X.; Wang, N.; Yan, N.; Shan, X.; Li, J. Preparation of γ-Divinyl-3-Aminopropyltriethoxysilane modified lignin and its application in flame retardant poly(lactic acid). *Materials* **2018**, *11*, 1505. [CrossRef]

18.   Mandlekar, N.; Cayla, A.; Rault, F.; Giraud, S.; Salaün, F.; Malucelli, G.; Guan, J. Thermal stability and fire retardant properties of Polyamide 11 microcomposites containing different lignins. *Ind. Eng. Chem. Res.* **2017**, *56*. [CrossRef]

19.   Mandlekar, N.; Malucelli, G.; Cayla, A.; Rault, F.; Giraud, S.; Salaün, F.; Guan, J. Fire retardant action of zinc phosphinate and polyamide 11 blend containing lignin as a carbon source. *Polym. Degrad. Stab.* **2018**, *153*, 63–74. [CrossRef]

20.   Mandlekar, N.; Cayla, A.; Rault, F.; Giraud, S.; Salaün, F.; Guan, J. Valorization of industrial lignin as biobased carbon source in fire retardant system for Polyamide 11 blends. *Polymers* **2019**, *11*, 180. [CrossRef]

21.   Levchik, S.V.; Costa, L.; Camino, G. Effect of the fire-retardant, ammonium polyphosphate, on the thermal decomposition of aliphatic polyamides. I. Polyamides 11 and 12. *Polym. Degrad. Stab.* **1992**, *36*, 31–41. [CrossRef]

22.   Troitzsch, J. *Plastics Flammability Handbook: Principles, Regulations, Testing, and Approval*, 3rd ed.; Carl Hanser Verlag: Munich, Germany, 2004; pp. 514–516.

23.   Lyon, R.E.; Walters, R.N. Pyrolysis combustion flow calorimetry. *J. Anal. Appl. Pyrolysis* **2004**, *71*, 27–46. [CrossRef]

24.   Brebu, M.; Vasile, C. Thermal Degradation of Lignin—A review. *Cellul. Chem. Technol.* **2010**, *44*, 353–363.

25. Rodrigues, J.; Graça, J.; Pereira, H. Influence of tree eccentric growth on syringyl/guaiacyl ratio in *Eucalyptus globulus* wood lignin assessed by analytical pyrolysis. *J. Anal. Appl. Pyrolysis* **2001**, *58–59*, 481–489. [CrossRef]
26. Wittkowski, R.; Ruther, J.; Drinda, H.; Rafiei-Taghanaki, F. Formation of smoke flavor compounds by thermal lignin degradation. In *Proceedings of the ACS Symposium Series*; American Chemical Society (ACS): Washington, DC, USA, 1992; Volume 490, pp. 232–243.
27. Kandola, B.K.; Horrocks, A.R. Complex char formation in flame-retarded fibre-intumescent combinations—II. Thermal analytical studies. *Polym. Degrad. Stab.* **1996**, *54*, 289–303. [CrossRef]
28. Wang, Y.; Zhang, F.; Chen, X.; Jin, Y.; Zhang, J. Burning and dripping behaviors of polymers under the UL94 verticla burning test conditions. *Fire Mater.* **2010**, *34*, 203–215. [CrossRef]
29. 29 Joseph, P.; Tretsiakova-McNally, S. Melt-flow behaviours of thermoplastic materials under fire conditions: Recent experimental studies and some theoretical approaches. *Materials* **2015**, *8*, 8793–8803. [CrossRef]
30. Dorez, G.; Ferry, L.; Sonnier, R.; Taguet, A.; Lopez-Cuesta, J.-M. Effect of cellulose, hemicellulose and lignin contents on pyrolysis and combustion of natural fibers. *J. Anal. Appl. Pyrolysis* **2014**, *107*, 323–331. [CrossRef]

Article

Lignin Nanoparticles as A Promising Way for Enhancing Lignin Flame Retardant Effect in Polylactide

Benjamin Chollet [1,2], José-Marie Lopez-Cuesta [1], Fouad Laoutid [2,*] and Laurent Ferry [1]

[1]   Centre des Matériaux des Mines d'Ales, IMT Mines Ales, University of Montpellier, 30319 Alès, France
[2]   Laboratory of Polymeric & Composite Materials, Materia Nova Research Center, 3 avenue Nicolas Copernic, B-7000 Mons, Belgium
*   Correspondence: fouad.laoutid@materianova.be

Received: 6 June 2019; Accepted: 28 June 2019; Published: 2 July 2019

Abstract: The present study investigates the effect of using lignin at nanoscale as new flame-retardant additive for polylactide (PLA). Lignin nanoparticles (LNP) were prepared from Kraft lignin microparticles (LMP) through a dissolution-precipitation process. Both micro and nano lignins were functionalized using diethyl chlorophosphate (LMP-diEtP and LNP-diEtP, respectively) and diethyl (2-(triethoxysilyl)ethyl) phosphonate (LMP-SiP and LNP-SiP, respectively) to enhance their flame-retardant effect in PLA. From the use of inductively coupled plasma (ICP) spectrometry, it can be considered that a large amount of phosphorus has been grafted onto the nanoparticles. It has been previously shown that blending lignin with PLA induces degradation of the polymer matrix. However, phosphorylated lignin nanoparticles seem to limit PLA degradation during melt processing and the nanocomposites were shown to be relatively thermally stable. Cone calorimeter tests revealed that the incorporation of untreated lignin, whatever its particle size, induced an increase in pHRR. Using phosphorylated lignin nanoparticles, especially those treated with diethyl (2-(triethoxysilyl)ethyl) phosphonate allows this negative effect to be overcome. Moreover, the pHRR is significantly reduced, even when only 5 wt% LNP-SiP is used.

Keywords: lignin nanoparticles; flame retardancy; polylactide; phosphorylation; biobased materials

---

1. Introduction

Biobased polymers have expanded significantly during the last decade. Many scientific studies have shown the ability to develop polymers from renewable resources (such as corn, wheat, sugar cane, etc.) in order to propose alternatives to fossil-based polymers, notably polylactic acid (PLA) [1,2]. Following the same trend, the development of green additives and fillers rouses more and more interest in order to reduce the environmental footprint of flame retarded polymers and particularly the biobased ones.

Studies about the development of fire retardants from biobased materials have been reviewed and widely discussed [3,4]. For example, the use of modified or unmodified polysaccharides (starch [5], chitosan [6]), proteins [7], or phenolic biomass [8] has been studied to improve the fire behavior of polymer materials. Among the phenolic biomass, lignin has been intensely studied [9–11] because of its abundance and its char forming ability that make it a good candidate for reducing the flammability of polymeric materials. Lignin can act as flame retardant for isotactic polypropylene at a relatively low incorporation rate (15 wt%) compared to the amount of fire retardant usually used in polypropylene [12]. Lignin may also be combined with some phosphate compounds and aluminum hydroxide to provide an increase in the thermal degradation temperature, the combustion time and the amount of residue from combustion of polypropylene. The compatibilization of lignin in acrylonitrile butadiene styrene has been investigated [13]. It has been shown that the incorporation of lignin can reduce the peak of heat release rate (pHRR), total heat release (THR), and mass loss rate during the combustion of acrylonitrile-butadiene styrene. The use of organosolv and Kraft lignins in PLA has been compared and the results demonstrated that blends properties depend on the nature of lignin employed [14]. Both lignins reduced the pHRR and the THR during cone calorimeter test owing to the formation of an insulating char layer at the surface of the sample during combustion. However, a decrease of the time to ignition (TTI) and the thermal stability of PLA was also observed.

At the same time, nano-technologies have allowed the improvement of the fire behavior of polymers owing to several processes such as the modification of the degradation pathway of the polymeric matrix as well as the modification of the rheological behavior. The thermal stability and flammability of poly(methyl methacrylate) (PMMA) containing organo-modified montmorillonite (OMMT) and metal oxide nanoparticles or both have been examined [15]. Thermal stability and flammability were improved with increasing amounts of oxide nanoparticles. The combination of oxide nanoparticles and organoclays induced a synergistic effect on the thermal stability and the fire performance: TTI was enhanced and THR was reduced. The combustion behavior of poly(ethylene-co-vinyl acetate) (EVA) composites containing modified organoclay has been studied and results showed that pHRR reduction was strongly affected by the nanoclay dispersion state [16]. All these studies highlight the superior flame-retardant effect of nanoparticles when properly dispersed in the polymeric matrix.

Only a few studies have associated both bio and nano-technologies for the development of flame retardants additives. The utility of this approach has been demonstrated especially when nanoparticles are associated with phosphorous compounds in the case of cellulose nanocrystals [17]. Lignin may be potentially appropriate for this approach since lignin nanoparticles can be easily obtained in different ways: Sonication [18], chemical modification [19], and precipitation [20,21]. However, lignin nanoparticles have never been used as flame retardants for polymeric materials. On the other hand, chemical functionalization of lignin with phosphorus-based molecules has been shown as an effective method for enhancing flame-retardant behavior [14,22,23]. Thus, the association of lignin nanoparticles with surface modification by grafting of phosphorous compounds is expected to entail a flame-retardant effect that could be effective at low levels of loading.

In this study, lignin nanoparticles (LNP) were prepared from Kraft lignin microparticles (LMP) by dissolution-precipitation. LMP and LNP were functionalized using diethyl chlorophosphate and diethyl (2-(triethoxysilyl)ethyl) phosphonate. All these lignins were incorporated into polylactic acid by melt blending. Thermal properties and fire behavior of the blends were determined using thermogravimetric analysis (TGA) and cone calorimeter experiments.

2. Materials and Methods

2.1. Materials

PLA resin (3052D) was purchased from NatureWorks (NatureWorks, Minnetonka, MN, USA). Kraft water soluble lignin with a low sulfonate content, diethyl chlorophosphate, and diethyl (2-(triethoxysilyl)ethyl) phosphonate were purchased from Sigma-Aldrich (Sigma-Aldrich, St Quentin Fallavier, France). Diethyl (2-(triethoxysilyl)ethyl) phosphonate was purchased from Specific Polymers (Specific Polymers, Castries, France). Absolute ethanol, ethylene glycol, and acetone were purchased from PanReac AppliChem (PanReac AppliChem, Darmstadt, Germany). Hydrochloric acid solution (35%) was purchased from Fisher Scientific (Fisher Scientific, Illkirch, France).

2.2. Lignin Nanoparticle Preparation

Lignin nanoparticles (LNP) were prepared from Kraft water-soluble lignin microparticles (LMP), by precipitation from ethylene glycol solution. A 4 wt% solution of lignin in ethylene glycol was prepared at 30 °C under stirring for 2 h. A 0.25M aqueous HCl solution was added at the rate of

2 drops per minute. After addition, the solution was stirred for another hour. Then, the solution was dialyzed against ethanol over 4 days in order to provoke non-solvent precipitation of lignin.

2.3. Lignin Treatment

With diethyl chlorophosphate (diEtP) (Figure 1a): 30 g of lignin in 800 mL of acetone were introduced in a two-necked round-bottom flask equipped with a refrigerant and heated at 60 °C. At reflux, 15 g of diethyl chlorophosphate were slowly added and the mixture was stirred for 5 h. After cooling, solution was washed 3 times with acetone by centrifugation. Functionalized LMP and LNP were noted LMP-diEtP and LNP-diEtP.

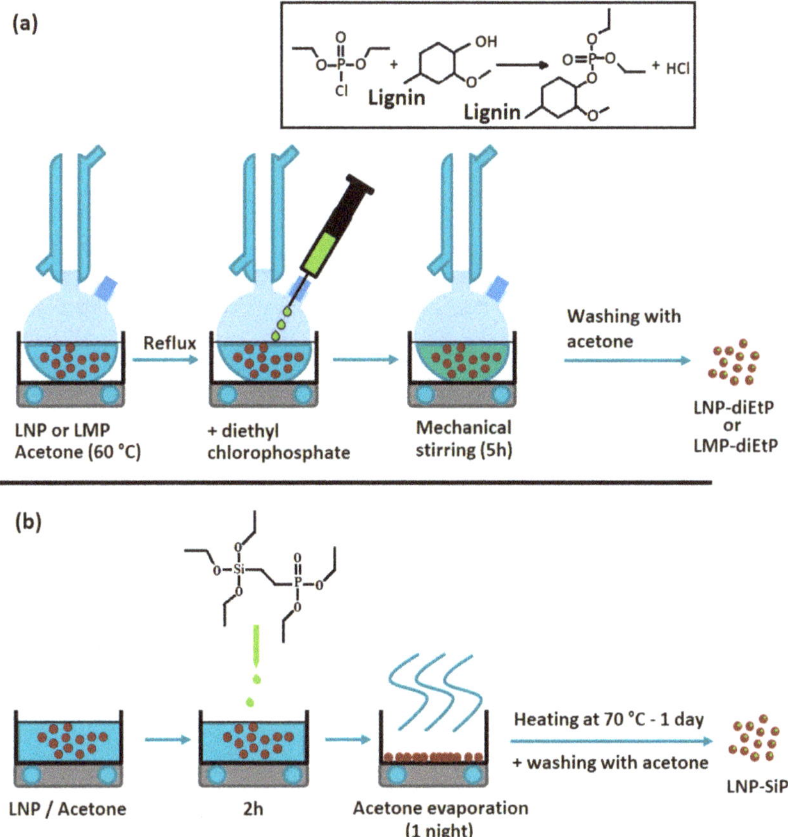

Figure 1. Schematic representation of lignin modification route used for grafting di-EtP (**a**) and SiP (**b**).

With diethyl (2-(triethoxysilyl) ethyl) phosphonate (SiP) (Figure 1b): a mixture of 30 g of lignin and 100 mL of acetone were introduced in 500 mL one neck round-bottom flask under mechanical stirring. 15 g of silane agent was slowly introduced. The mixture was stirred over 2 h at room temperature and first dried under a ventilated hood over 1 day and at 70 °C for 1 night. The obtained dry powders have been washed 3 times with acetone by centrifugation. Functionalized LMP and LNP were noted LMP-SiP and LNP-SiP, respectively.

2.4. Composites Preparation

PLA 3052D containing 5 wt% and 10 wt% of the untreated and treated lignin micro and nano-particles were prepared in a Rheomix 3000 internal mixer from Thermo Fisher Scientific. PLA was first ground in 2 mm diameter powder and a suspension of lignin particles in acetone was added to the powder. Acetone was evaporated under a ventilated hood, then the powder mixture was dried overnight in a ventilated oven at 70 °C. Powders were blended in the internal mixer at 170 °C, first at 30 rpm during 3 min and then at 70 rpm during 7 min. The obtained composite was ground into 6 mm diameter pellets. Composite plates of $100 \times 100 \times 3$ mm^3 were compression molded from pellets using a Darragon thermopress at 170 °C. First pellets were brought into contact with the heated platens for 4 min, then a 50 bar pressure was applied for 2 min and finally a 100 bar pressure was applied for 2 more min. Plates were cooled in ambient atmosphere.

2.5. Lignin and Composites Characterization

a- Particle Size Analysis

Laser particle size analyzer LS 13 320 from Beckman-Coulter Company was used for determining particle size distribution of untreated and treated lignin nanoparticles in acetone. Particle size distribution was obtained by scattering of monochromatic light with a wavelength of 780 nm diffracted and transmitted through the suspension. Each analysis was at least reproduced 3 times and the maximum relative standard deviation for the median size was 10%. Quanta 200 FEG Scanning Electron Microscope (SEM) from FEI Company (FEI, Northeast Dawson, Hillsboro, OR, USA) was also used for investigating the lignin particles shape under high vacuum at a voltage of 12.5 kV and a working distance between 8.2 mm and 10.6 mm.

b- Phosphorus Content

Phosphorus content in functionalized lignins was determined using an Activa M Inductively Coupled Plasma (ICP) spectrometer from Horiba Jobin Yvon (HORIBA FRANCE SAS, Montpellier, France). In a first step, the organic matrix was decomposed by a mineralization process using HNO_3 and H_2SO_4 solutions, followed by microwaves irradiation using a Milestone 1200-Mega from Gemini BV (Gemini BV, Apeldoorn, the Netherlands). A calibration curve was used to determine the exact amount of phosphorus in the samples.

c- Thermal Degradation

Thermal degradation of lignins and composites was studied by thermogravimetric analysis (TGA). Dried lignins samples and composites were submitted to a temperature ramp from 70 °C to 700 °C at a heating rate of 10 °C/min. TGA were performed under both air and nitrogen flow of 100 mL/min using a Setsys Evolution device from Setaram Instrumentation. 5%, 10%, and 50% weight loss temperatures (respectively $T_{5\%}$, $T_{10\%}$, and $T_{50\%}$) and char yield at 650 °C were determined.

d- Fire Properties

The fire behavior of the different composites was studied using a cone calorimeter from Fire Testing Technology. Samples of $100 \times 100 \times 3$ mm^3 were exposed to a 35 kW.m^{-2} radiant heat flux, corresponding to common heat flux in a mild fire scenario. Heat Release Rate (HRR) was measured as a function of time and Time To Ignition (TTI) and peak of Heat Release Rate (pHRR) were determined. For each sample, at least 2 tests were performed.

e- Degradation of PLA During Melt Processing

PLA thermal degradation during melt processing was investigated using a Steric Exclusion Chromatography (SEC) analysis. Samples were dissolved in chloroform (2 mg.mL^{-1}) and filtered through a 0.2 µm filter. The molecular weight distributions were determined in $CHCl_3$ at 23 °C using an Agilent size exclusion chromatograph equipped with a Knauer 2320 refractometer index detector and two PLGel columns (MIXED-D and 103 A). 20 µL of the sample solutions were injected into the columns using a flow rate of 1 mL/min. Monodisperse polystyrene standards (Polymer Laboratories Ltd., Church Stretton, UK) were used for the primary calibration. The number average molecular weight (Mn) was determined.

3. Results

3.1. Lignin Particles Structural and Thermal Properties

3.1.1. Unmodified Lignin

First, the process of the lignin nanoparticles preparation was assessed. SEM observations (Figure 2) and particle size analyses (Figure 3) were used to evaluate the evolution of both particles size and their shape. It is important to notice that these analyses were performed on dry powders. Both laser particles size analyses and SEM observations evidenced that the original Kraft lignin contains microparticles with diameter ranging from 5 µm up to some hundred microns. Particles obtained using the dissolution-precipitation process exhibit different aspects. In fact, SEM micrographs highlight an important change of the powder morphology (Figure 2) that seems more like a porous and spongy material composed of the aggregation of small nanoparticles. Determination of particle size distribution supports this observation and shows an important change of the grain-size distribution curve that highlights the formation of very small nanoparticles with a d_{50} and d_{90} of only 80 nm and 160 nm, respectively.

Thermogravimetric analyses (TGA) were performed under both air and nitrogen to evaluate the effect of lignin particle size on its thermal stability (Figure 4). The thermal degradation of lignin is widely described in the literature [24]. Its decomposition starts by a water release followed by the first decomposition step (230–260 °C) that leads to the formation of low molecular weight products due to propanoid side chain cleavage. The main decomposition step occurs at higher temperature (250–450 °C) and leads to the production of a large quantity of methane due to the cleavage of lignin main chain and followed (above 500 °C) by several rearrangements and condensation reactions of the aromatic structure that leads to the formation of char structures, which decompose above 650 °C. This char is thermally stable only under anaerobic condition and decomposes totally in the presence of oxygen. TGA curves, under air and under nitrogen of LMP particles (Figure 4) follow the general decomposition steps described above and only the decomposition step occurring above 800 °C, corresponding to the thermo-oxidative decomposition of the char is affected by the nature of the atmosphere. In contrast, it is interesting to notice that the reduction of particle size strongly affect lignin thermo-oxidative degradation. Under air, the weight loss recorded above 350 °C is significantly more important in the case of LNP, while under pyrolytic conditions, only a slight difference between the curves of LMP and LNP was observed above 350 °C. This difference could not be attributed to lignin chemical structure since both particles were obtained from the same original kraft lignin and only particle size changed. The presence of oxygen played an important role for promoting this degradation. As LNP sample was composed of a porous material (aggregates of nanoparticles), the oxygen diffusion through this material during TGA is expected to be easier with respect to more compact micronic LMP particles. LMP were thus less exposed to oxygen since they present a lower specific surface area.

Figure 2. SEM micrographs of untreated and treated lignin microparticles (LMP) and lignin nanoparticles (LNP).

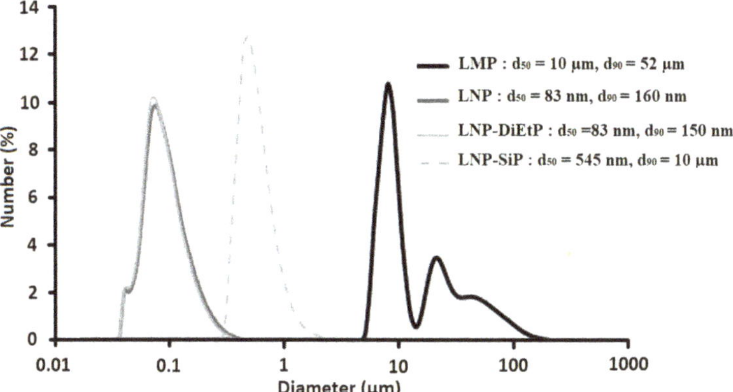

Figure 3. Number particle size distribution for LMP, LNP, and LNP-diEtP determined by laser particle size analyses in acetone.

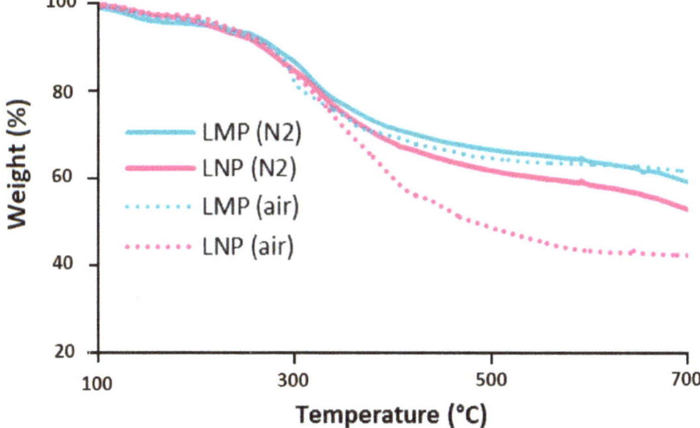

Figure 4. TGA curves, under air and nitrogen for LMP and LNP (10 °C/min).

3.1.2. Phosphorylated Lignin

The phosphorylation process is mainly used to increase the char stability during lignin aerobic thermal degradation and thus enhances the flame-retardant effect of lignin. First of all, the amount of grafted phosphorus with both diethyl chlorophosphate and diethyl (2-(triethoxysilyl) ethyl) phosphonate has been determined. ICP analyses (Table 1) highlight that, whatever the nature of grafting agent, phosphorus content is always higher when lignin nanoparticles are used, and the highest phosphorus content was obtained when lignin nanoparticles are treated with diethyl (2-(triethoxysilyl) ethyl) phosphonate (3.2%). Using multifunctional silyl-based phosphonate enables obtaining higher P content. This result was expected since diethyl chlorophosphate is able to react with only one lignin hydroxyl group. In contrast, SiP could establish 3 bonds with 3 hydroxyl groups provided by lignin or by self-condensation. However, it is important to notice that P content obtained is very low (only 0.1 wt%) when SiP is used with LMP. This result reflects the competition between grafting reaction on lignin and self-condensation. In fact, if only a few hydroxyl groups are present on the surface of lignin particles, the self-condensation between SiP molecules will consume the main part of the reactive functions of the modifying agent at the expense of grafting reactions. This result also highlights that the following cleaning procedure enables the elimination of ungrafted molecules.

Table 1. Phosphorus content determined by ICP of the different treated lignin particles.

Sample	Phosphorus Content (wt%)
LMP-diEtP	0.35 ± 0.02
LNP-diEtP	0.67 ± 0.02
LMP-SiP	0.1 ± 0.02
LNP-SiP	3.2 ± 0.02

However, grafting SiP induces some changes on the lignin particles structural and thermal properties. Firstly, in contrast to diEtP that does not affect lignin particle size, using SiP induces a significant increase in lignin particle size. Indeed, Figure 3 shows that lignin nanoparticles modified with diethyl chlorophosphate exhibit similar size compared to unmodified nanoparticles, while the size of lignin nanoparticles increases from 83 nm (d_{50}) and 150 nm (d_{90}) for LNP to 545 nm (d_{50}) and 10 μm (d_{90}) for LNP-SiP. The increase of lignin particle size observed in the case of LNP treated with SiP results from self-condensation of SiP as well as from the coupling of different nanoparticles due to the multifunctional reactivity of SiP. Such behavior has also been observed in the literature and resulted in an increase of lignin molecular weight [17] and lignin particle size [19]. SEM observations performed on dry LNP-SiP (Figure 2) support this result since the morphology of LNP-SiP particles changes and aggregates formed from small particles are evidenced. Moreover, EDX analysis of these aggregates reveals the presence of high concentrations of Si and P. It is important to notice that PLA / lignin nanocomposites were prepared by mixing PLA powders with a suspension of lignin nanoparticles into acetone in order to avoid the aggregation of lignin nanoparticles during the drying process. Hence, particle size analyses carried out in acetone provide a more accurate information upon the particle size and the particle size determined from SEM images is strongly dependent on the drying process that induces particle aggregation.

All these observations enable us to propose the following schematic representation of LNP-SiP (Figure 5) that explains the origin of the high P content as well as the significant increase of particles size in the case of LNP modified with SiP.

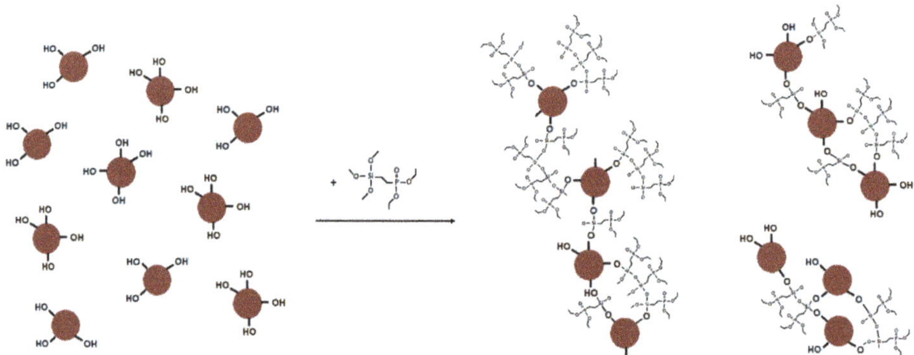

Figure 5. Schematic representation of LNP-SiP particles.

The effect of lignin phosphorylation on its thermal stability has been also evaluated using TGA analysis (Figure 6). The phosphorylation process of lignin is designed to increase the thermo-oxidative resistance of the char but mainly this reaction induces its premature thermal decomposition [10]. In our case, TGA analysis performed under nitrogen does not show any effect of the phosphorylation process on either the thermal stability or the amount of the final residue.

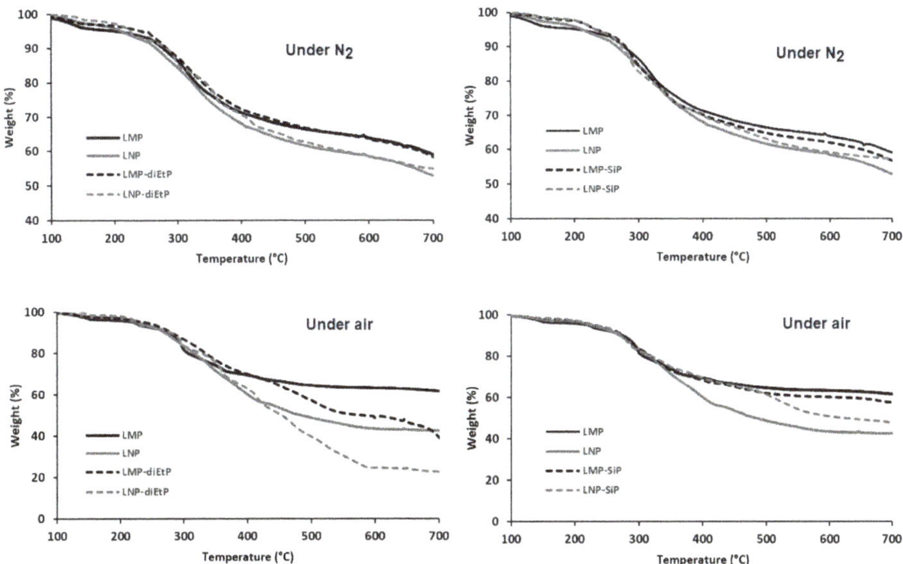

Figure 6. TGA curves of untreated and phosphorylated lignin particles (under air and nitrogen, 10 °C/min).

On the contrary, under air, the presence of phosphorus induces important changes in TGA curves (Figure 6) and the thermal behavior of modified lignins is significantly affected by the nature of the phosphorous agent used. In the case of lignin treated with diEtP, a significant reduction of the thermal stability of both modified lignin particles was observed above 400 °C. Instead of the plateau observed between 400 °C and 700 °C for both unmodified lignin particles, which corresponds to the formation of thermally stable char residues, the weight loss still occurs gradually above 400 °C in the case of LMP-diEtP and LNP-diEtP. It is worth to mention that the final residues at 700 °C are lower in the case of treated lignins, whatever their particle size. The effect of phosphorus is surprising since it was expected to act as a char promoter and not to induce char degradation. In the presence of SiP agent, the thermal behavior of treated and untreated LMP particles is not significantly affected. This result is due to the low content of grafted SiP. However, in the case of LNP, the effect of surface modification is more significant, and an important enhancement of the char thermal stability is observed above 400 °C. In fact, the degradation of the char previously observed with diEtP is strongly limited and the phosphorous agent acts mainly as a stabilizing agent. The amount of the final char at 700 °C is slightly higher for SiP treated LNP, but not in the case of treated microparticles.

3.2. Composites Thermal Degradation and Fire Behavior

It has been proven that lignin could be advantageously used to improve the char formation during the combustion of polymeric matrices [13,14,17]. However, all works reported in literature studied the effect of lignin microparticles and none of them concerned the effect of lignin nanoparticles. At least 20 wt% of lignin was considered, in order to significantly improve the composite fire behavior. Using lignin at nanoscale is expected to enable reducing its incorporation rate, while maintaining its flame-retardant effect owing to "the nano" effect [25]. In fact, generally speaking, nanoparticles, when correctly dispersed, are well known to reduce pHRR during cone calorimeter test.

3.2.1. Effect of Untreated Lignin Content and Particle Size

First, we investigated the effect of lignin particle size on PLA degradation during melt processing, thermal stability by TGA and fire properties using cone calorimeter test. Both lignin micro and nano-particles were incorporated in PLA by melt mixing at different contents (5 and 10 wt%). SEM observation in Figure 7 evidences a rather good LNP dispersion since different nanoparticles of about 100 nm are observed as well as some aggregates of about 5 μm.

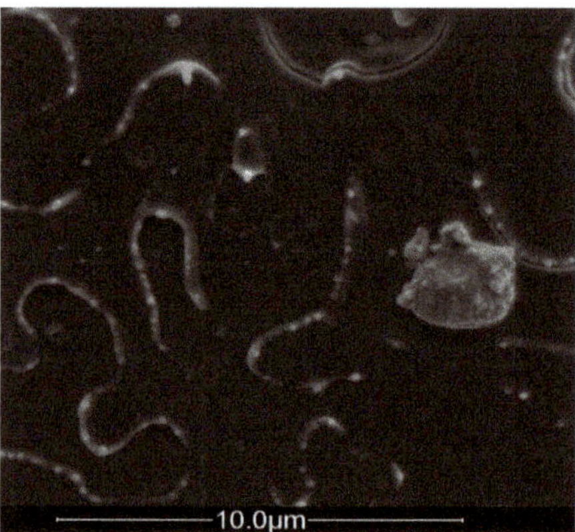

Figure 7. SEM image of PLA containing 5wt% LNP (White dots represent LNP).

It has been reported in a previous work [17] that the incorporation of lignin into PLA presents some drawbacks: PLA degradation occurs during melt processing due to the presence of some degradant groups such as phenolic and carboxylic functions, as well as sulfur groups present at the surface of lignin particles.

Thus, the effect of the incorporation of both lignins (nanoparticles and microparticles) on the thermal degradation of PLA during melt processing was first studied to evaluate whether the incorporation of very small lignin particles, which are supposed to exhibit higher specific surface area (1 m^2/g for LMP and around 70 m^2/g for LNP theoretically), induces higher degradation with respect to lignin microparticles. Figure 8 and Table 2 shows the evolution of the number average molecular weight (Mn) of processed PLA containing different lignin contents. The results indicate that increasing lignin content, whatever the particle size, induces significant reduction of Mn. The incorporation of 10 wt% lignin particles induces a reduction of about 50% of PLA molecular weight that decreases from 80,000 g/mol to 40,000 g/mol. Using 20 wt% LMP or LNP induces further reduction of Mn up to 20,000 g/mol. The PLA composite presenting such a low molecular weight will exhibit very low mechanical properties. To overcome this limitation, it is important to reduce the lignin content.

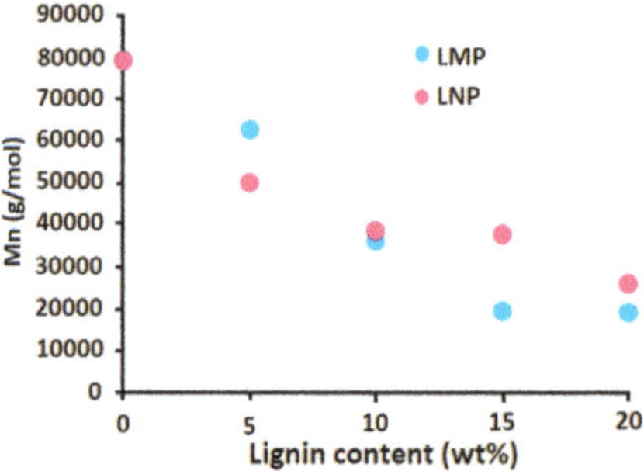

Figure 8. Effect of lignin size and content on the evolution of the number average molecular weight of processed PLA.

Increasing lignin content, whatever the particle size, also induces an important reduction of PLA thermal stability during TGA analysis under nitrogen (Figure 9). Using lignin nanoparticles does not induce further degradation and the TGA curve of the composite containing 20 wt% lignin remains slightly similar whatever the particle size. Similar reduction of PLA thermal stability in the presence of lignin has been also evidenced for PLA containing 20 wt% lignin [14,26] and was attributed to the action of degradant lignin phenolic hydroxyl and carboxylic groups. Using lignin nanoparticles, that present higher surface area does not induce any further reduction of PLA thermal stability with respect to microparticles, since PLA degradation is induced by degradant groups formed during lignin thermal decomposition and not by those present at the surface. The latter are only effective at low temperature and mainly induce PLA degradation during melt processing.

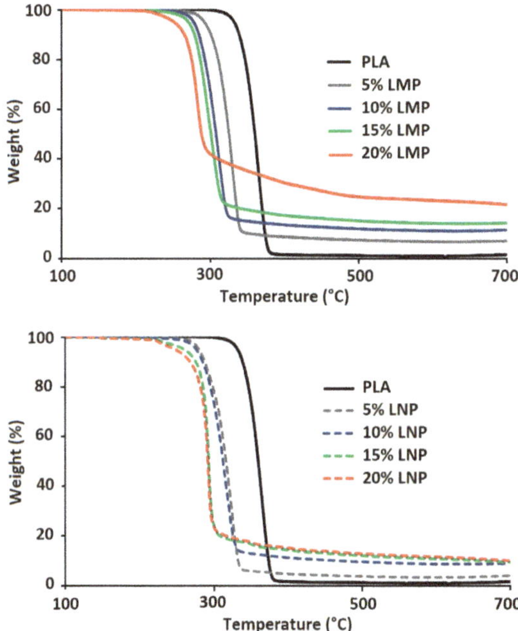

Figure 9. Effect of lignin size and content of PLA thermal stability during TGA under nitrogen.

The incorporation of untreated lignins entails some modifications of the fire behavior of PLA assessed using cone calorimeter test at 35 kW/m^2. Figure 10 shows pHRR curves of PLA containing different lignin contents and Figure 11 summarizes these results. In fact, both LMP and LNP induce important reduction of the resistance to ignition of PLA. Reduction of TTI in the presence of lignin has already been reported in the literature about PLA [26], PBS [22], and ABS [23]. It was attributed to several factors such as polymer thermal degradation, modification of composites emissivity, heat absorption and thermal conductivity. However, the reduction of PLA composites resistance to ignition is higher in the presence of nanoparticles and therefore lower TTI are observed (Figure 11). Using lignin nanoparticle thus induces important changes in physical and physico-chemical properties of the composite that significantly affects the composites resistance to ignition. The reason behind this effect is not clearly established but may be induced by an increase of nanocomposite heat absorption that could promote a quicker heating and the rapid formation of combustible volatile products. Further studies are needed for a better understanding of this effect.

Moreover, the incorporation of lignin particles, whatever their size, does not induce any significant reduction of pHRR except the composition containing 20 wt% LMP that only presents however a slight reduction of about 10%. In addition, using reduced lignin content does not present any pHRR reduction. Using such a low lignin content (5 wt%) is not enough to promote the formation of a continuous protective layer and only an 'islands-in-the-sea' structure is generated.

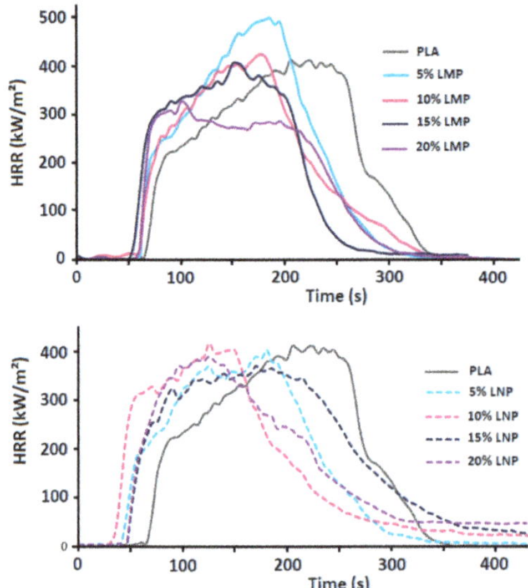

Figure 10. HRR curves of PLA containing different LMP and LNP content.

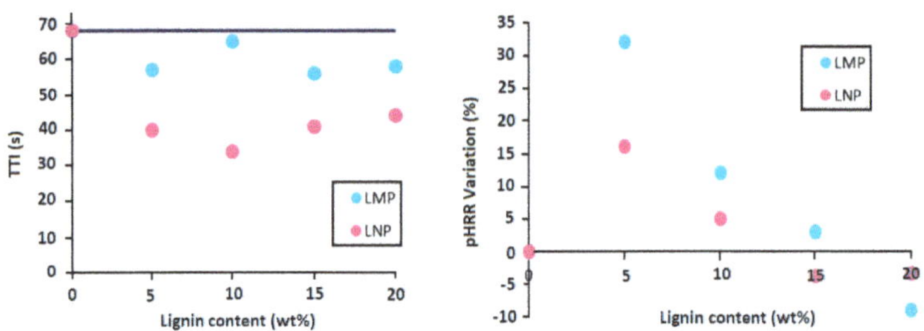

Figure 11. Evolution of time to ignition (TTI) and pHRR variation of PLA containing 5; 10; 15 and 20 wt% LMP and LNP.

The amount of the final char left after cone calorimeter test is mainly affected by lignin particle size (Figure 12). Using lignin nanoparticles induces the formation of lower char residue at the end of the test because of some incandescence phenomena that induce char degradation after the flame out.

Fire testing results clearly indicate that no superior effect was obtained using untreated lignin nanoparticles. Performances remain similar to those obtained with microparticles.

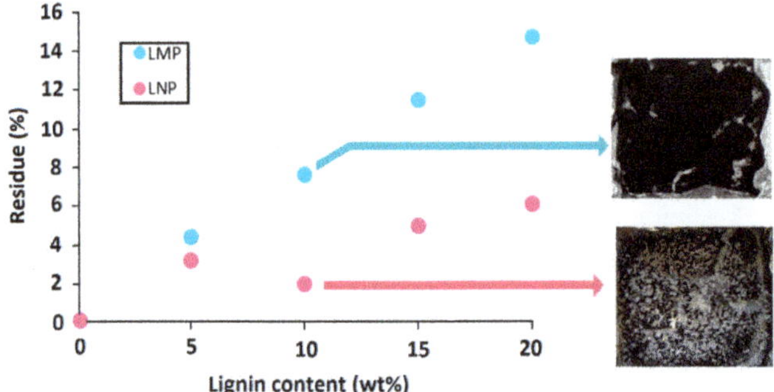

Figure 12. Effect of lignin size and content on the amount of char residue formed during cone calorimeter test.

3.2.2. Effect of Phosphorylated Lignin

The results obtained with untreated lignin nanoparticles did not highlight any superior effect of LNP with respect to LMP. The other way for taking advantage of lignin nanoparticles is to modify their surface by phosphorous based compound in order to limit their degradant effect, enhancing the amount of the char as well as its thermal stability for developing nanocomposites presenting high molecular weight and improved fire performance. With this view, untreated and phosphorylated lignin micro and nano particles were incorporated into PLA at reduced content (5 and 10 wt%). Their effect on PLA degradation, its thermal stability and fire behavior were studied.

Phosphorylation of lignin nanoparticles doesn't enable significant increase of PLA thermal stability during melt processing. In fact, PLA molecular weights remain similar to those obtained with untreated LMP and LNP. Some changes are observed but do not demonstrate a significant effect and the Mn remains lower to 55,000 g/mol, while PLA containing 10 wt% untreated lignin microparticles and nanoparticles present Mn around 40,000 g/mol. The effect resulting from the presence of LNP-SiP is a little bit disappointing regarding the presence of a significant amount of non-degradant SiP groups at the surface of lignin. The number of average molecular weight of PLA composites containing LNP-SiP and LMP-SiP are very similar to those obtained with untreated LNP and LMP, respectively. The presence of SiP seems to not affect PLA thermal degradation during melt processing.

Table 2. Number average molecular weight of processed PLA and PLA composites.

	Mn (g/mol)
PLA	80,000
10 wt% LMP	36,000
10 wt% LNP	39,000
10 wt% LMP-diEtP	33,000
10 wt% LNP-diEtP	55,000
10 wt% LMP-SiP	30,000
10 wt% LNP-SiP	40,000

SEC analysis enables determining the degradant effect of additives on PLA during melt processing at 170 °C while TGA analysis gives information on composites thermal stability at higher temperatures. TGA tests were run on dried samples, with a ramp of temperature from 100 °C to 700 °C. From these analyses, temperatures corresponding to 5% weight loss ($T_{5\%}$), 10% weight loss ($T_{10\%}$), 50% weight loss ($T_{50\%}$), as well as the value of the final char at 650 °C, were determined. TGA curves are presented in Figure 13 and data summarized in Table 3.

The addition of untreated lignin, whatever its particle size, induced an important reduction of PLA thermal stability for all compositions. $T_{5\%}$ and $T_{10\%}$ were reduced, respectively, from 333 °C and 340 °C for neat PLA to below 290 °C and 300 °C in the presence of lignin particles. Following the same trend, $T_{50\%}$ was decreased from 360 °C to below 326 °C. Figure 14 shows that some correlation exists between $T_{5\%}$ and PLA chain molecular weight. The more Mn was reduced and the lower was $T_{5\%}$. This observation has already been reported in the literature [14]. As previously mentioned, this could be attributed to the degradant effect of some moieties present at the lignin surface and seems to indicate that the thermal degradation occurring during melt processing continued to occur at higher temperature during TGA experiments. Another hypothesis could be that thermal stability of PLA depends on the molecular weight of its macromolecules [27]. It means that shorter polymer chains would be more easily thermally degraded than larger polymer chains. It is worth mentioning that the use of nanoparticles did not induce any further degradation since $T_{5\%}$ and $T_{10\%}$ are similar to those obtained with 10 wt% LMP. These results showed that lignin particle size did not induce any significant modification of PLA thermal stability. Moreover, the incorporation of both lignins also induced the formation of some char during thermal decomposition under both air and nitrogen. However, this char was only thermally stable under anaerobic conditions. In fact, under air, almost no residue (maximum 1.3%) was left at the end of the test. At similar incorporation content, LMP enabled obtaining higher char yield than LNP. This could be due to the volatilization of a part of lignin nanoparticles with the degradation product. Moreover, under nitrogen, experimental char yields of the composites filled with LMP were higher than theoretical ones. This result evidences the protective effect of lignin that, owing to its char forming ability, avoids total PLA degradation, leading to a higher residue. With respect to untreated lignins, the incorporation of chemically treated lignin particles into PLA entailed different changes on composites thermal stability. While the presence of phosphorylated microparticles induced the early degradation of PLA under nitrogen, the incorporation of LNP-SiP nanoparticles enabled maintaining the thermal stability of PLA since the nanocomposite TGA curve was very close to that of pristine PLA. These results clearly indicate the interest of using phosphorylated nanoparticles instead of microparticles and especially those treated by SiP. The superior effect of LNP-SiP nanoparticles is related to their higher phosphorus content.

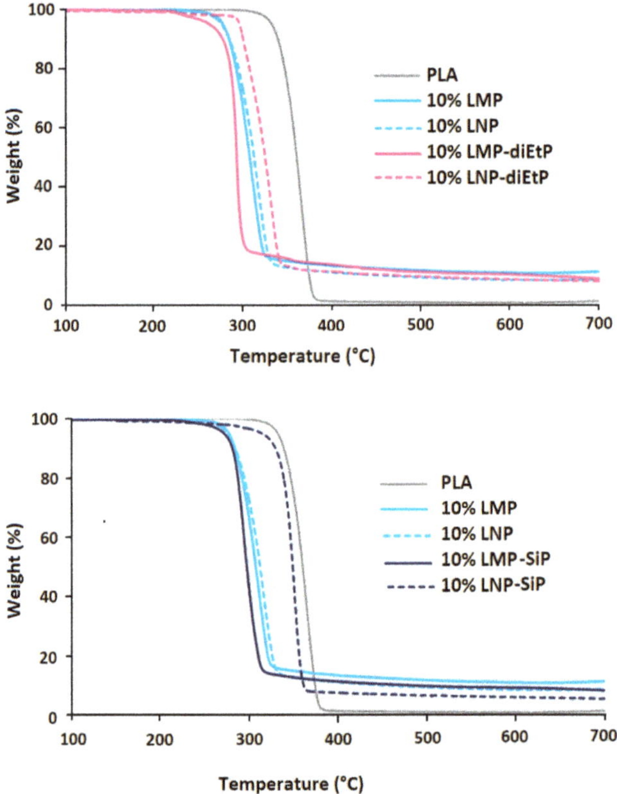

Figure 13. TGA curves of PLA and PLA composites under nitrogen.

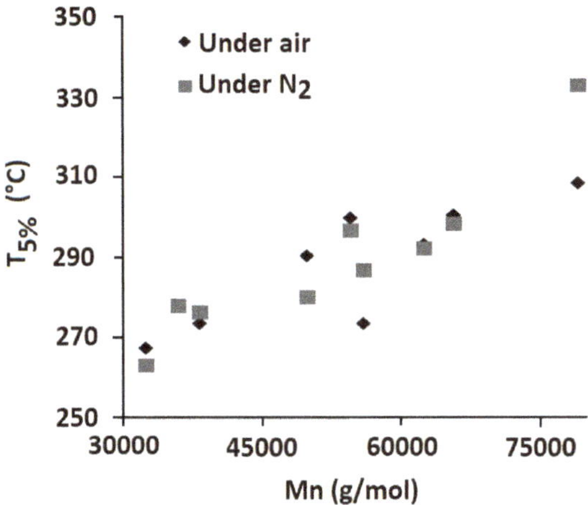

Figure 14. T5% as function of Mn.

Table 3. TGA results of neat PLA and related composites under nitrogen (10 °C/min).

	$T_{5\%}$ (°C)	$T_{10\%}$ (°C)	$T_{50\%}$ (°C)	Residue at 650 °C (%)
PLA	333	340	360	0.9
10% LMP	278	285	308	10.8
10% LNP	277	285	313	8.5
10% LMP-diEtP	263	278	294	9.7
10% LNP-diEtP	297	300	325	8.3
10% LMP-SiP	273	282	300	8.8
10% LNP-SiP	315	330	350	5.7

The effect of the incorporation of phosphorylated lignin nanoparticles on the composites fire behavior was investigated using cone calorimeter tests. As can be seen in Figure 15 and Table 4, the phosphorylation of lignin nanoparticles using diEtP enables limiting the reduction of TTI observed when untreated nanoparticles are used. Hence, time to ignition increases from 34 s obtained with 10 wt% LNP to 54 s when LNP-diEtP are used. This is likely due to the increase of the composite thermal stability due to the grafting of non-degradant diethylphosphate groups on the degradant hydroxyl groups, as was attested by both SEC and TGA analysis. However, using phosphorylated lignin with diEtP, whatever its particle size, does not enable any reduction of pHRR that remains at least similar to that of pristine PLA. The limited flame-retardant effect of these lignins seems to be mainly governed by the low phosphorus content grafted when diEtP is used.

However, using SiP as grafting agent enables obtaining higher performances with treated lignin nanoparticles (Figure 16). Hence, the incorporation of 10 wt% LNP-SiP entails an important increase of time to ignition that reaches 86 s and exceeds thus the time to ignition of pristine PLA. This result confirms the interest of using LNP-SiP that was shown to maintain the thermal stability of PLA as noted from TGA analysis. In addition, using 10 wt% LNP-SiP enables a significant reduction of pHRR of about 18%. This result is very important since this it is the first time that is reported that using so reduced lignin content (only 10 wt%) enables such significant flame-retardant effects, i.e., TTI increase and pHRR reduction. SEM observations and EDX analysis evidence a homogeneous dispersion of LNP-SiP in the composite (Figure 17). In fact, different EDX analyses were performed at different zones and most of them evidence the presence of both P and Si. It is important to notice that SEM analyses do not evidence the presence of numerous large particles aggregates in the composites, but only few of them are visible. The major LNP-SiP particles are properly dispersed.

Interestingly enough, the flame retardant effect of LNP-SiP remains effective even at more reduced content. Indeed, the incorporation of only 5 wt% LNP-SiP (Figure 16) still allows for obtaining high time to ignition (84 s) and reduced pHRR (−11%). Despite those noticeable effects, the amount of the final residue is low (1%). This result was assumed to be due to the incandescence phenomenon that induces char degradation after the flame out.

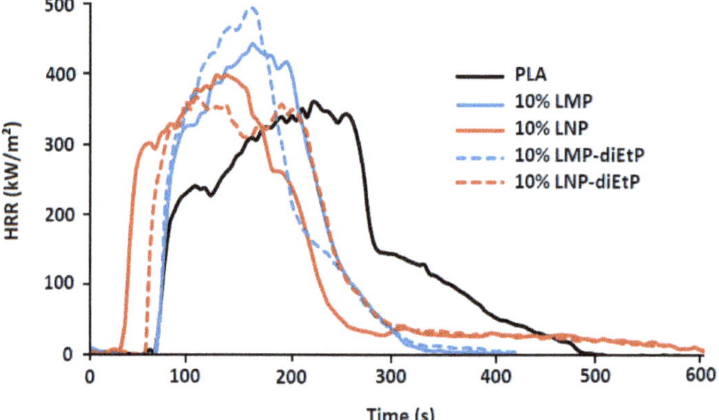

Figure 15. HRR curves of PLA during cone calorimeter test.

Table 4. Data of cone calorimeter test.

	TTI (s)	pHRR Variation (%)	Char (%)
PLA	68	/	0.1
10% LMP	65 ± 4	+12 ± 2	11.5 ± 0.4
10% LNP	34 ± 2	+5 ± 2	2 ± 0.1
10% LMP-diEtP	67 ± 2	+20 ± 8	8 ± 1.2
10% LNP-diEtP	54 ± 2	−6 ± 1	4.2 ± 1.3
10% LMP SiP	57 ± 4	+20 ± 5	7.3 ± 0.3
10% LNP SiP	86 ± 7	−18 ± 1	2 ± 4
5% LMP-SiP	57 ± 3	+ 18 ± 2	4.5 ± 0.2
5% LNP-SiP	84 ± 4	−11 ± 2	1 ± 0.1

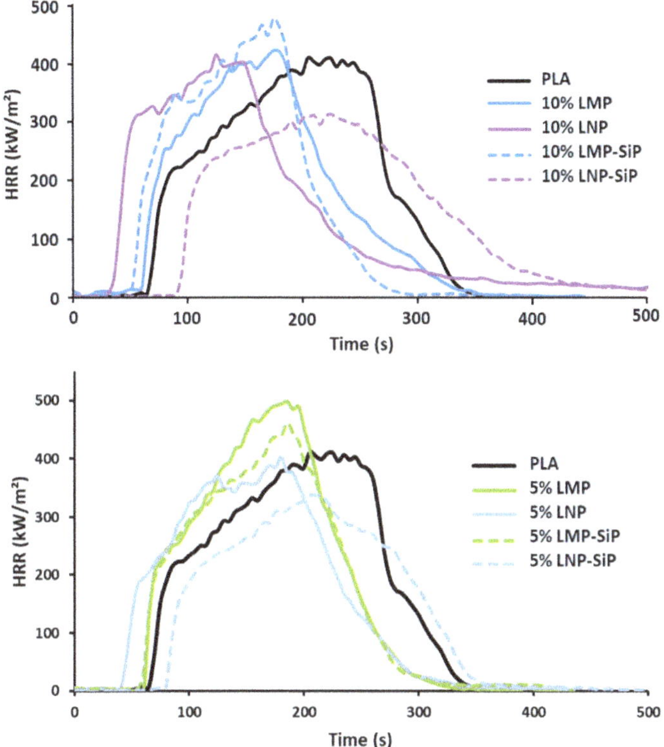

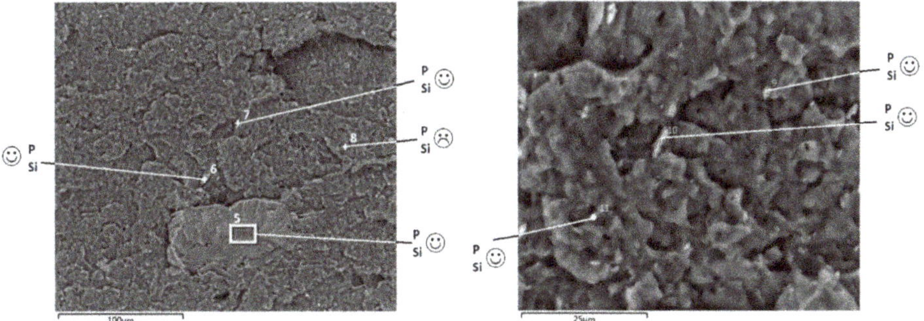

Figure 16. HRR curves of PLA during cone calorimeter test.

Figure 17. SEM observations and EDX analysis of PLA containing 10 wt% LNP-SiP.

4. Conclusions

Lignin nanoparticles were synthesized from Kraft lignin. Both micro and nano lignins were functionalized with diethyl chlorophosphate and diethyl (2-(triethoxysilyl) ethyl) phosphonate. Using diethyl (2-(triethoxysilyl) ethyl) phosphonate enables grafting higher P content. Moreover, highest P content was obtained with lignin nanoparticles (3.2 wt%). While both untreated lignin micro and nano-particles induce dramatic reduction of PLA thermal stability during TGA analysis, grafting SiP enables limiting this negative effect and enhancing the composites thermal stability that becomes very close to that of unfilled PLA. While the other lignin particles do not show any flame-retardant effect into PLA, the incorporation of LNP-SiP content at relatively low content (10 wt%) into PLA enables important increase of time to ignition as well as a reduction of pHRR. Remarkably enough, this effect is obtained even with the lowest LNP-SiP content (5 wt%). This study reports, for the first time, the interest of using lignin nanoparticles as flame-retardant agent in polymers. Results showed that grafting significant P content at the surface of lignin nanoparticles enables their use as flame-retardant additives, effective even at low incorporation content (5 wt%).

Author Contributions: Conceptualization, F.L., L.F. and J.-M.L.-C.; investigation, B.C.; resources and project administration, F.L., L.F.; data curation, writing—original draft preparation and visualization, B.C.; validation, writing—review and editing, F.L., L.F. and J.-M.L.-C.; supervision, F.L. and L.F.

Funding: This research was funded by IMT Mines Ales and Materia Nova.

Conflicts of Interest: The authors declare no conflict of interest.

References

1. Farah, S.; Anderson, D.G.; Langer, R. Physical and mechanical properties of PLA, and their functions in widespread applications—A comprehensive review. *Adv. Drug Deliv. Rev.* **2016**, *107*, 367–392. [CrossRef] [PubMed]

2. Murariu, M.; Dubois, P. PLA composites: From production to properties. *Adv. Drug Deliv. Rev.* **2016**, *107*, 17–46. [CrossRef] [PubMed]

3. Costes, L.; Laoutid, F.; Brohez, S.; Dubois, P. Bio-based flame retardants: When nature meets fire protection. *Mater. Sci. Eng. R Rep.* **2017**, *117*, 1–25. [CrossRef]

4. Sonnier, R.; Taguet, A.; Ferry, L.; Lopez-Cuesta, J.M. *Towards Bio-Based Flame Retardant Polymers*; Springer Briefs in Molecular Science; Springer International Publishing: Cham, Switzerland, 2018.

5. Sakharov, A.M.; Sakharov, P.A.; Lomakin, S.M.; Zaikov, G.E. Chapter 7—Novel Class of Eco-Flame Retardants Based on the Renewable Raw Materials. In *Polymer Green Flame Retardants*; Elsevier: Amsterdam, The Netherlands, 2014; pp. 255–266.

6. Hu, S.; Song, L.; Pan, H.; Hu, Y.; Gong, X. Thermal properties and combustion behaviors of flame retarded epoxy acrylate with a chitosan based flame retardant containing phosphorus and acrylate structure. *J. Anal. Appl. Pyrolysis* **2012**, *97*, 109–115. [CrossRef]

7. Carosio, F.; di Blasio, A.; Cuttica, F.; Alongi, J.; Malucelli, G. Flame retardancy of polyester and polyester–cotton blends treated with caseins. *Ind. Eng. Chem. Res.* **2014**, *53*, 3917–3923. [CrossRef]

8. Howell, B.; Oberdorfer, K.L.; Ostrander, E.A. Phosphorus Flame Retardants for Polymeric Materials from Gallic Acid and Other Naturally Occurring Multihydroxybenzoic Acids. *Int. J. Polym. Sci.* **2018**, *3*, 1–12. [CrossRef]

9. Dehne, L.; Vila Babarro, C.; Saake, B.; Schwarz, K.U. Influence of lignin source and esterification on properties of lignin-polyethylene blends. *Ind. Crops Prod.* **2016**, *86*, 320–328. [CrossRef]

10. Laurichesse, S.; Avérous, L. Chemical modification of lignins: Towards biobased polymers. *Prog. Polym. Sci.* **2014**, *39*, 1266–1290. [CrossRef]

11. Naseem, A.; Tabasum, S.; Zia, K.M.; Zuber, M.; Ali, M.; Noreen, A. Lignin-derivatives based polymers, blends and composites: A review. *Int. J. Biol. Macromol.* **2016**, *93*, 296–313. [CrossRef]

12. De Chirico, A.; Armanini, M.; Chini, P.; Cioccolo, G.; Provasoli, F.; Audisio, G. Flame retardants for polypropylene based on lignin. *Polym. Degrad. Stab.* **2003**, *79*, 139–145. [CrossRef]

13. Song, P.; Cao, Z.; Fu, S.; Fang, Z.; Wu, Q.; Ye, J. Thermal degradation and flame retardancy properties of ABS/lignin: Effects of lignin content and reactive compatibilization. *Thermochim. Acta* **2011**, *518*, 59–65. [CrossRef]

14. Costes, L.; Laoutid, F.; Aguedo, M.; Richel, A.; Brohez, S.; Delvosalle, C.; Dubois, P. Phosphorus and nitrogen derivatization as efficient route for improvement of lignin flame retardant action in PLA. *Eur. Polym. J.* **2016**, *84*, 652–667. [CrossRef]

15. Laachachi, A.; Leroy, E.; Cochez, M.; Ferriol, M.; Lopez Cuesta, J.M. Use of oxide nanoparticles and organoclays to improve thermal stability and fire retardancy of poly(methyl methacrylate). *Polym. Degrad. Stab.* **2005**, *89*, 344–352. [CrossRef]

16. Zanetti, M.; Kashiwagi, T.; Falqui, L.; Camino, G. Cone calorimeter combustion and gasification studies of polymer layered silicate nanocomposites. *Chem. Mater.* **2002**, *14*, 881–887. [CrossRef]

17. Costes, L.; Laoutid, F.; Khelifa, F.; Rose, G.; Brohez, S.; Delvosalle, C.; Dubois, P. Cellulose/phosphorus combinations for sustainable fire retarded polylactide. *Eur. Polym. J.* **2016**, *74*, 218–228. [CrossRef]
18. Gilca, I.A.; Popa, V.I.; Crestini, C. Obtaining lignin nanoparticles by sonication. *Ultrason. Sonochem.* **2015**, *23*, 369–375. [CrossRef] [PubMed]
19. Gilca, I.A.; Ghitescu, R.E.; Puitel, A.C.; Popa, V.I. Preparation of lignin nanoparticles by chemical modification. *Iran. Polym. J.* **2014**, *23*, 355–363. [CrossRef]
20. Frangville, C.; Rutkevičius, M.; Richter, A.P.; Velev, O.D.; Stoyanov, S.D.; Paunov, V.N. Fabrication of environmentally biodegradable lignin nanoparticles. *ChemPhysChem* **2012**, *13*, 4235–4243. [CrossRef]
21. Yang, W.; Kenny, J.M.; Puglia, D. Structure and properties of biodegradable wheat gluten bionanocomposites containing lignin nanoparticles. *Ind. Crops Prod.* **2015**, *74*, 348–356. [CrossRef]
22. Ferry, L.; Dorez, G.; Taguet, A.; Otazaghine, B.; Lopez-Cuesta, J.M. Chemical modification of lignin by phosphorus molecules to improve the fire behavior of polybutylene succinate. *Polym. Degrad. Stab.* **2015**, *113*, 135–143. [CrossRef]
23. Prieur, B.; Meub, M.; Wittemann, M.; Klein, R.; Bellayer, S.; Fontaine, G.; Bourbigot, S. Phosphorylation of lignin to flame retard acrylonitrile butadiene styrene (ABS). *Polym. Degrad. Stab.* **2016**, *127*, 32–43. [CrossRef]
24. Brebu, M.; Vasile, C. Thermal degradation of lignin—A review. *Cellul. Chem. Technol.* **2010**, *44*, 353–363.
25. Laoutid, F.; Bonnaud, L.; Alexandre, M.; Lopez-Cuesta, J.-M.; Dubois, P. New prospects in flame retardant polymer materials: From fundamentals to nanocomposites. *Mater. Sci. Eng. R Rep.* **2009**, *63*, 100–125. [CrossRef]
26. Costes, L.; Laoutid, F.; Brohez, S.; Delvosalle, C.; Dubois, P. Phytic acid-lignin combination: A simple and efficient route for enhancing thermal and flame retardant properties of polylactide. *Eur. Polym. J.* **2017**, *94*, 270–285. [CrossRef]
27. Ray, S.; Cooney, R.P. Thermal degradation of polymer and polymer composites. In *Handbook of Environmental Degradation of Materials*; Elsevier: Amsterdam, The Netherlands, 2012; pp. 213–242.

Article

Preparation of a Novel Flame Retardant Formulation for Cotton Fabric

Hung Kim Nguyen [1], Wataru Sakai [2] and Congtranh Nguyen [1,*]

[1] Department of Polymer Chemistry, Faculty of Chemistry, University of Science, Viet Nam National University of Ho Chi Minh City, Ho Chi Minh City 700000, Vietnam; nkhung@hcmus.edu.vn
[2] Faculty of Materials Science and Engineering, Kyoto Institute of Technology, Matsugasaki, Sakyo, Kyoto 606-8585, Japan; wsakai@kit.ac.jp
* Correspondence: nctranh@hcmus.edu.vn; Tel.: +84-983-353-354

Received: 29 October 2019; Accepted: 6 December 2019; Published: 20 December 2019

Abstract: A novel halogen-free flame-retardant formulation was prepared and coated onto cotton fabrics. The structure of phosphorus compounds in the system was characterized by attenuated total reflectance Fourier transform infrared spectroscopy (ATR-FTIR) and nuclear magnetic resonance spectroscopy (^{1}H-NMR). Results from the ATR-FTIR spectroscopy, scanning electron microscopy (SEM), and energy-dispersive X-ray spectroscopy (EDX) analyses presented that the flame retardant was coated successfully onto a cotton surface. We investigated the thermal stability and fire-retardant behaviors of cotton fabrics using thermal gravimetric analysis (TGA) and the vertical flame test. We also discuss the mechanism of flame retardance of coated cotton fabrics.

Keywords: organophosphorus compounds; flame retardant; cotton fabrics; condensed phase

---

1. Introduction

Cotton fabrics are known as one of the most important natural fibers found in apparel production, home furnishings, and industrial product applications. This is due to cotton's excellent properties such as biodegradability, softness, warmth, and recyclability [1]. However, the high flammability of cotton fabrics is a consideration in cotton-based productions [1,2]. A proper flame retardant (FR) should be applied to retard the ignition of the cotton fabrics and/or decrease flame spread [02]. Among the various technological methods developed by academics to prepare the durable FR finishing of cotton fabrics, FR coatings have become one of the most convenient, economical, and most efficient ways [3]. Cotton fabrics are often treated with reactive FR or they are back coated to improve flame retardancy behavior. For example, silicones [4,5], polyurethanes [6], and polyphosphonate (PEPBP) [7,8] coated onto cotton surfaces are utilized to provide adequate flame retardancy under adverse environmental conditions. In addition to polymer matrix-based FR, various active FRs were proposed to impart flame retardancy to cotton fabrics [9]. Given environmental concerns related to halogen FRs, phosphorus-based FRs are considered the most promising candidate because of their ecological properties and fire retardance in both the gas phase and condensed phase [10–12]. The recent researches have found the condensed-phase effectiveness in the preparation of phosphorus-based FR coatings or back coated for the cotton fabric process [13–16]. The FR compounds above may also contain phosphorus elements or combinations of P and N elements for synergistic interaction. The cross-linkable organophosphorus FR system was introduced into cotton fabrics. From the cotton or cotton/nylon blends treated with commercial FRs, oligomeric OH terminated methylphosphonate-phosphate has exhibited a highe char residue due to the catalysis effect of phosphoric acid on the cotton/nylon fabrics for the dehydration of cellulose. The char length of the treated cotton or treated cotton/nylon blends decreases compared to the length of the untreated cotton or single-fiber nylon fabrics [17].

Recently, multifunctional FR monomers containing triazine rings [18,19] were synthesized and introduced onto the cotton fabrics. Triazine-based and UV-curable flame retardants were also integrated as alternative durable FRs for cotton fabrics through polymerization under UV radiation [20]. C–O bonding was formed through active chlorine atoms in the triazine-based FRs owing to the hydroxyl bonding in the cellulose units. It was found that the durability of flame resistance of cotton fabric that diminished with triazine-based FRs may be improved significantly after some soaping cycles.

Other phosphorus-based FRs reported in the literature include phosphorus-containing acrylates, such as methacryloloxyethylorthophosphor tetraethyl diamidate (MPD), pyrovatex, tri(acryloyloxyethyl) phosphate, and triglycidyl isocyanurate acrylates. Acrylate FRs are introduced to the substrate through the UV curing or impregnation method. It was found that a chemical cross-linking reaction occurred between the FRs and cotton matrix, revealing excellent thermal stability under high-temperature conditions. Moreover, the FR behavior of the treated cotton indicated higher initial degradation temperature and more char yield compared to that of the neat cotton.

To develop novel phosphorus-containing FRs for cotton fabrics, this study synthesized a new FR formulation using anhydride methacrylic and bis(hydroxymethyl)phosphinic acid. The formulation's structure was characterized using FTIR and ^{1}H NMR analysis. The surface and thermal behaviors of the treated cotton were characterized using ATR-FTIR, SEM, EDX, and TGA analysis. Furthermore, we proposed the FR mechanism of the organophosphorus formulation on cotton fabrics by investigating the chemical structure and elemental compositions of the char residues.

2. Experimental

2.1. Materials

Cotton fabric: 100% scoured and bleached plain-weave cotton fabrics were purchased from commercial entities in Vietnam with a density of 237 g/m^2.

Methacrylic anhydride (MAAH) was purchased from Sigma-Aldrich Co. Munich, Germany. Benzoyl peroxide (BPO) was purchased from Merck Co. Darmstadt, Germany. Triethylamine (TEA), acetone, 37% HCl solution, and 25% NH$_3$ solution were purchased from Xilong Chemical Co., shantou, China. All reagents were used without any further purification.

2.2. Preparation of the Phosphorus Containing Formulation

Bis(hydroxymethyl)phosphinic- methacrylate (PMA) system was synthesized through the two-step reaction presented in Scheme 1.

Starting bis(hydroxymethyl)phosphinic acid (BHMP) was not commercial, and the authors prepared it according to the protocol described in Reference [21]. BHMP (10.08 g, 0.08 mol) and triethylamine (8.08 g, 0.08 mol) were placed into a three neck-round-bottom flask equipped with a temperature controller, a reflux condenser, and a mechanical stirrer. The temperature of the flask was maintained at 0–5 °C. Methacrylic anhydride (24.64 g, 0.16 mol) was placed in an addition funnel, and it was added slowly to the flask containing the BHMP solution over an hour. Then, the reaction mixture was stirred at 40 °C for 24 h. A yellowish product was obtained with a yield of 90%.

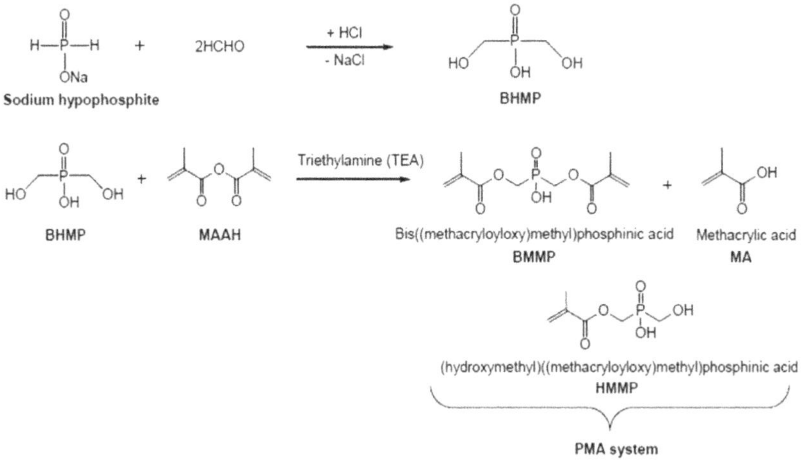

Scheme 1. Synthesis route of PMA system.

2.3. Fabric Treatment

For irradiation, we used a metal halide lamp (made in China) with a broad-band UV-source. Blank cotton fabrics were immersed in the acetone solution containing the FR system (as shown in Table 1) in the presence of 1% BPO at room temperature and then neutralized by ammonia solution to pH 7–8. After 30 min, the impregnated fabric was removed from the solution, placed on a glass plate, dried at 90 °C for 3 min, and then cured by UV radiation on both sides for 5 min each. After radiation, the fabric was soaked with water and then sequence impregnated in 0.5% HCl solution and 1% NH_3 solution for 3 min. Finally, after washing ten times, each sample was allowed to air dry at room temperature (30 °C) until no weight loss was detected.

Table 1. Formulations, weight gain of samples.

Sample	Composition FR(g)/Acetone (L)	Weight Gain after Washing 10 Times (%)
COT-20	380	20.73
COT-25	440	25.70
COT-30	500	30.27

2.4. Measurements

2.4.1. FTIR and ^{1}H-NMR

FTIR spectroscopy was carried out on a Jasco FT/IR 4700, Kyoto, Japan to characterize the structure of the PMA using a thin KBr disk. The surface functional groups of the samples were investigated by ATR-FTIR using a diamond crystal at 32 scans (Jasco FT/IR 4700). The measurement was carried out in the range of 4000–500 cm^{-1} by 32 scans, where the resolution was 4 cm^{-1}.

The ^{1}H-NMR measurement was recorded on a Bruker AV 500 MHz spectrometer, Ho Chi Minh City, Vietnam, in DMSO, with tetramethylsilane (TMS) as the reference.

2.4.2. Weight Gain

All cotton fabrics were dried in an oven at 90 °C for 30 min, and then weighed quickly. The weight gain of the treated fabric was obtained using the following equation:

$$\text{Weight gain (\%)} = 100 \times (W_2 - W_1)/W_1 \tag{1}$$

where W_1 and W_2 represent the weights of the untreated fabrics and treated ones, respectively.

2.4.3. Scanning Electron Microscopy (SEM)

The surface morphology and chemical compositions of the fabric samples were acquired using SEM (Hitachi S-3000N, Kyoto, Japan) equipped with EDX spectrometers. The specimens has been coated with a conductive layer.

2.4.4. Vertical Flame Test

The vertical flame test was carried out according to the DIN 53906 standard method [22]. The sample size was 150 mm × 75 mm. Butane gas was selected for the combustion. The flame height and burning time were about 40 mm and 10 s, respectively. The average flaming time (for both the after flame and afterglow) of the five test specimens was recorded.

2.4.5. Thermogravimetric Analysis (TGA)

To characterize the thermal properties of the cotton fabrics, we employed thermogravimetric analysis (Discovery TA Instrument, Kyoto, Japan) under nitrogen and air atmospheres from 30 to 600 °C at a heating rate of 10 °C /min.

2.4.6. Durability Test

The flame durability test was evaluated using the ISO 105-C10:2006 standard [23]. The laundering process used a non-ionic surfactant. The temperature of the laundering solution was kept at approximately 40 °C. Each cotton sample was washed ten times continuously.

3. Results and Discussion

3.1. Synthesis of the PMA

A novel FR formulation was prepared using the two-step reaction presented in Scheme 1. Figure 1 shows the FTIR spectrum of the phosphorus-containing formulations. The absorption bands of the saturation carbon (sp^3) were observed at 2988–2900 and 1430 cm^{-1}. The peak at 3363 cm^{-1} contributed to the stretching vibration of the –OH group. The peaks at 1720 and 1626 cm^{-1} corresponded to the vibration of the C=O and C=C groups, respectively. The peaks at 1168 and 1041 cm^{-1} could be assigned to the stretching band of P=O and P–O–C bonding, respectively.

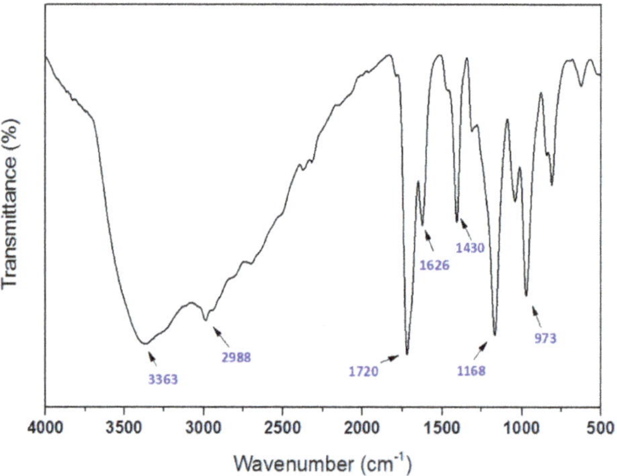

Figure 1. FTIR spectrum of PMA.

Figure 2 shows the ^{1}H-NMR spectrum of the phosphorus-containing formulation. The chemical shifts of protons were observed at 1.80–1.86 (s, CH$_3$–C), 3,59 (d, P–CH$_2$–OH), 4.20–4.25 (d, P–CH$_2$–OC(O)), and 5.60–6.05 (d, CH$_2$=CH–). The observed signals at 4.20 (d) and 4.25 (d), respectively, corresponded to the chemical shifts of the protons in P–CH$_2$ in the BMMP. Table 2 shows the peak assignments of three UV curable monomers (BMMP, HMMP, and MA) indicated in the FR formulation. The FTIR and ^{1}H–NMR results indicated that the novel FR formulation was prepared successfully, and it was able to be used as a UV curable FR formulation without further purification.

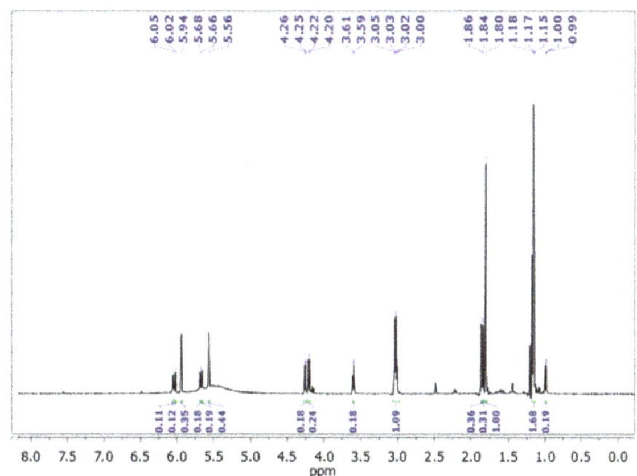

Figure 2. Nuclear magnetic resonance spectroscopy (^{1}H-NMR) spectrum of PMA.

Table 2. Proton signals of components in the FR formulation.

Substance	Signal	Group
BMMP	+ 1.86 (s) ppm + 4.20 (d) ppm + 5.66 (s) and 6.02 (s) ppm	C–CH$_3$ P–CH$_2$–OC(O) C=CH$_2$
HMMP	+ 1.84 (s) ppm + 3.59 (d) ppm + 4.25 (d) ppm + 5.68 (s) and 6.05 (s) ppm	C–CH$_3$ P–CH$_2$–OH P–CH$_2$–OC(O) C=CH$_2$
MA	+ 1.80 (s) ppm + 5.94 (s) and 5.56 (s) ppm	C–CH$_3$ C=CH$_2$

3.2. Surface Characterization of the Treated Cotton Fabrics

We investigated the fabric surface properties using ATR-FTIR and SEM image analyses. Figure 3 shows the ATR-FTIR spectrum of untreated (COT) and FR coated cotton fabrics (COT-30). The results showed that both COT and COT-30 fabrics displayed stretching vibration modes of –OH, C–H (sp^3), and C–O groups in cellulose at 3300, 2950, and 1050 cm^{-1}, respectively. Compared to the ART-FTIR spectrum of the uncoated COT, two new absorption appearances at 1710 cm^{-1} and 1546 cm^{-1} (Figure 3b) were attributed to –COO–CH$_2$–P (ester) and –COO– (carboxylate anion of the ammonium salt), respectively. These results concealed that the FRs were coated on the cotton surfaces.

The surface morphology structure of untreated and PMA-treated cotton fabrics was also characterized using SEM analysis. The smooth fabric surface was observed in the SEM image of the neat cotton fabrics (Figure 4A,A1). Meanwhile, the PMA-treated cotton fabrics displayed rougher and more intact features with many layers of polymer infiltrating to the cotton fabrics (Figure 4B–D). We observed that a lot of the PMA polymers covered the fibers in the COT-20, COT-25, and COT-30 fabrics. Denser layers were observed as the loadings of the FR increased.

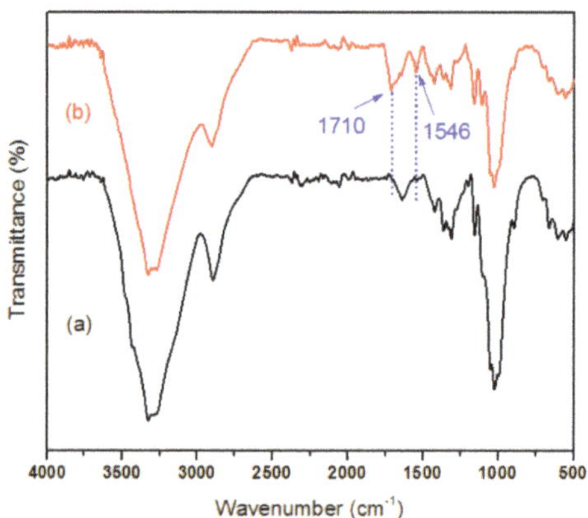

Figure 3. Attenuated total reflectance Fourier transform infrared spectroscopy (ATR-FTIR) spectrum of COT (**a**) and COT-30 (**b**).

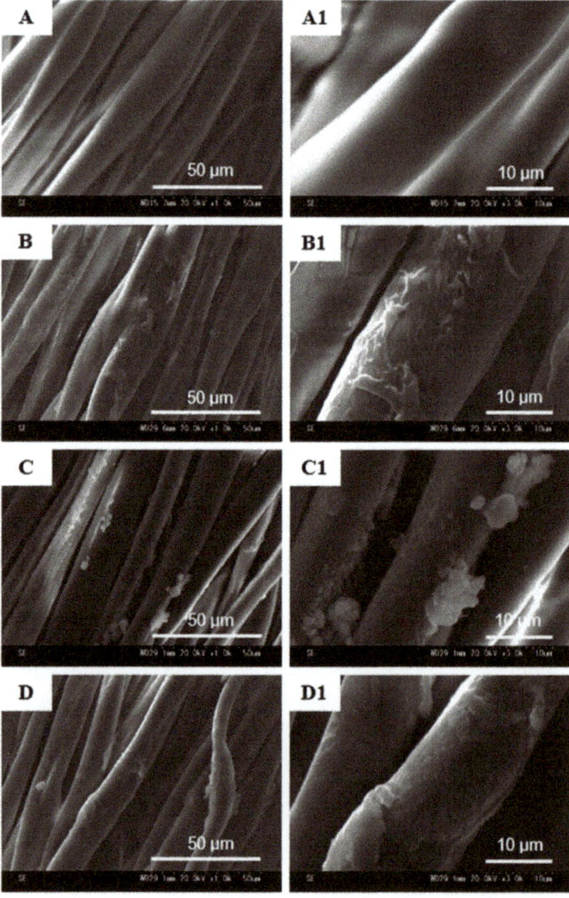

Figure 4. Scanning electron microscopy (SEM) images of COT (**A**,**A1**); COT-20 (**B**,**B1**); COT-25 (**C**,**C1**); and COT-30 (**D**,**D1**) at ×1000 and ×3000 magnification, respectively.

To further support the above assertion, the elemental compositions of the untreated and treated cotton fabrics were analyzed using EDX experiments. Figure 5 shows the EDX results of the COT, COT-20, COT-25, and COT-30 samples. We found there was limited P content (0.04% only) and low N content (3.67%) due to the non-cellulose impurities showing in fibers [01] in the COT sample. We found that changes in the P and N contents were determined from 0.23% to 0.60% and 6.06% to 6.50% as the weight loadings of FRs added increased from 20% to 30%, respectively. The noticeable increase in the P and N contents of the treated cotton fabrics was revealed compared to that of the untreated cotton fabrics. Moreover, the P and N contents on the treated cotton fabrics increased with an increasing PMA concentration in the finishing solution. From Figure 6, the data from the P and N element mapping images of the treated cotton fabrics showed that the P and N elements were almost uniformly distributed in all the treated cotton fabrics. The higher the weight loadings of the FR, the denser the density of distribution. Despite all the PMA-treated cotton fabrics being washed ten times with the soaping solution, the FRs were still retained on the cotton layers in all the PMA-treated cotton fabrics. Therefore, we concluded from these findings that the FR formulation was coated onto the surface of the cotton fabrics.

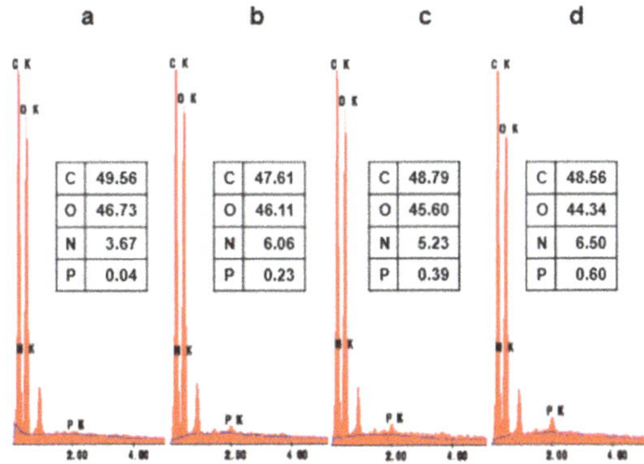

Figure 5. EDX results of COT (**a**), COT-20 (**b**), COT-25 (**c**), and COT-30 (**d**).

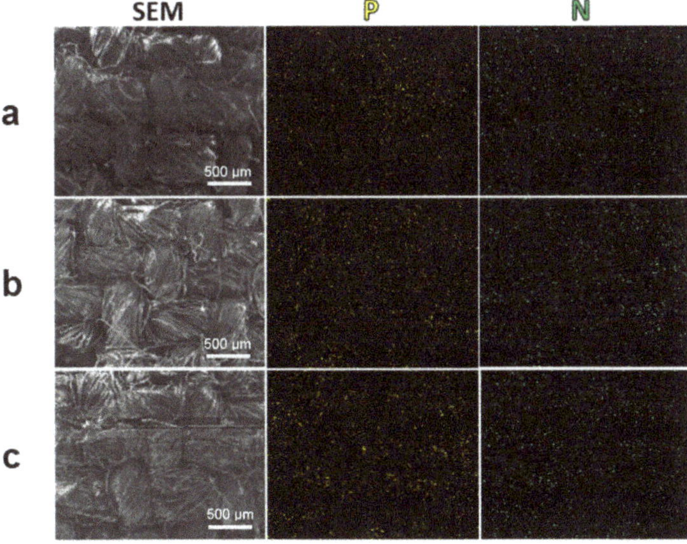

Figure 6. Mapping of the P and N elements in COT-20 (**a**), COT-25 (**b**), and COT-30 (**c**).

3.3. Flame Retardant Performance

We investigated the flame retarding performances of the treated cotton fabrics using the vertical flame test. Following DIN 53906, we recorded the average times for both the after flame and afterglow of each of the five test samples. The flammability results of the cotton fabrics and images taken after the vertical burning tests are shown in Table 3. The neat cotton fabrics were natural to burn and were almost destroyed without any char residues remaining. The COT sample burned out violently with after flame and afterglow times of 18 s and 14 s, respectively. On the contrary, the COT-A fabrics without phosphorous-containing FRs behaved as highly flammable fabrics, and no char residues were obtained after the combustion. However, a very thin and light char was formed during the COT-A combustion. This showed that ammonium carboxylate decomposition generated ammonia gas and carboxylic acid groups during the flaming. These carboxylic acid groups accelerated the dehydration of the cotton fabrics. We observed that increasingly dehydrated cotton formed thermally stable carbon-rich residues. The ammoniac release gas might dilute the vapor phase gas on the cotton surface. On the other hand, this diluted vapor phase gas partially retarded the complete degradation of the cotton into CO_2 and H_2O, though negligible [24].

Table 3. The test residue and vertical flame test results of the neat cotton and treated cotton fabrics.

Sample	COT	COT-20	COT-25	COT-30	COT-A

After flame time (s)	18	25	0	0	18
After glow time (s)	14	0	0	0	5
Char length (mm)	150	150	100	90	150
DIN 53906	Fail	Fail	Pass	Pass	Fail

Table 3 shows that the inflammability behaviors of the treated cotton fabrics were enhanced as more FRs were added. We noted that both the after time and afterglow times of the coated COT-fabrics were achieved at lower values compared to the values of the neat cotton or COT-A. Both COT-25 and COT-30 fabrics showed self-extinguishing with no after time and afterglow time observed after applying the 10 s burning test. In particular, both COT-25 and COT-30 fabrics exhibited char length lower than 150 mm, thereby passing the DIN 53906 standard for both COT-25 and COT-30 fabrics. Meanwhile, the higher after flame time of COT-20 compared to that of neat cotton was 25 s, while the afterglow time of the COT-20 fabrics was zero. This result indicated that the char formed from COT-20 burning was thermally stable. Since the char length reached over 150 mm, the COT-20 sample did not pass the DIN 53906 standard. It is considered that the char forming ability of UV curable BMMP and HMMP FRs were responsible for the improvement in flame retardancy on the cotton fabrics.

To understand the flame retardancy behaviors of the UV curable FRs, we investigated the thermal stabilities of the neat cotton, coated cotton fabrics, and PMA homopolymer using TGA analysis. All the TGA curves and thermal analysis data of the samples carried out under both nitrogen and oxygen atmosphere are shown in Figure 7, Figure 8, and Table 4, respectively. All TGA curves exhibited at least two main stages as the temperature increased from 50 °C to 600 °C with a heating rate of 10 °C /min. The 5 wt% of weight loss around 100–120 °C for all samples in the first stage corresponded to the evaporation of water due to moisture absorption. The primary decomposition stage with the highest-rated and highest weight loss (87%) at 330–400 °C was attributed to hydration and further cellulose degradation in the cotton [25]. For the FR coated cotton fabrics, we observed the same thermal degradation at a similar temperature range (260–350 °C) even though the weight loss was not significant (45–50%) compared to that of the neat cotton fabrics. The degradation appearing at lower temperatures (about 260 °C for FR coated COT samples and 300 °C for PMA homopolymer) was related to the formation of phosphoric acid derivatives from the decomposition of the FR, the cleavage of the aliphatic fraction, and the cellulose dehydration of the cotton fabrics [25]. The gradual degradation of both the FR coated cotton samples and PMA homopolymer appearing at 340–430 °C was related to network formation through the esterification of phosphinic acids. A large amount of charred residue was observed for all the FR coated cottons (about 27.0, 27.8, 30.1, and 35.2% in PMA homopolymer, COT-20, COT-25, and COT-30, respectively), while very little char (7.1%) was recorded in the neat cotton (Figure 7). Another noticeable thing was that all TGA curves of the samples in an oxygen atmosphere showed the same degradation behavior beyond 300 °C compared to the curves in the nitrogen atmosphere (Figure 9). However, a lower amount of char residues in the oxygen gas was obtained at 600 °C for both the neat cotton fabrics and FR coated cotton fabrics compared to the residues in the nitrogen medium. That is, 0.2 % for the COT and COT-A, and about 27.0, 27.8, 30.1, and 35.2% in PMA homopolymer, COT-20, COT-25, and COT-30, respectively. The lower amount resulted from these char residues being partially volatilized because of the further oxidation. With little char residue during the COT-A combustion, polymethacrylic coated cotton fabrics showed no contribution to enhancing the thermal stability behavior of the cotton fabrics. On the contrary, the formation of significant char residues through the further reaction of cotton with the phosphorus-containing FRs (COT-20, COT-25, and COT-30) was responsible for the flame inhibition action. Therefore, we concluded that the compound containing phosphinic moiety played a role in the efficient char-forming FR of cotton fabrics in the condensed phase. However, its initial degradation occurred at quite lower temperatures.

Table 4. Thermogravimetric analysis data of the samples.

Sample	Nitrogen		Oxygen	
	Onset (°C)	wt% of Char at 600°C	Onset (°C)	wt% of Char at 600°C
PMA	320	30.0	316	35.4
COT	343	7.1	331	0.2
COT-20	299	27.8	299	8.7
COT-25	294	30.1	293	12.4
COT-30	283	35.2	285	14.1
COT-A	326	8.0	310	0.1

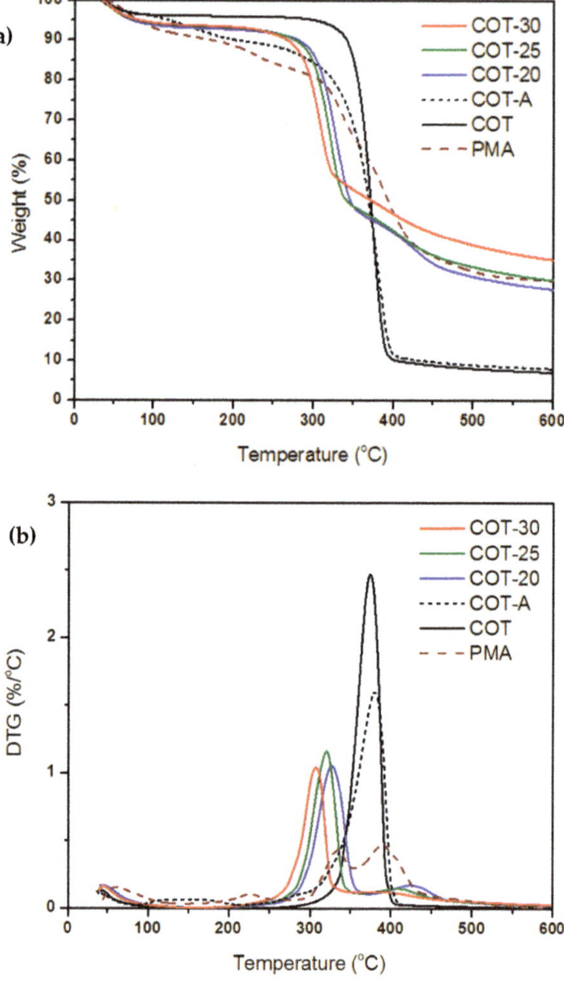

Figure 7. TGA (**a**) and DTG (**b**) curves of the neat cotton and treated cotton fabrics in a nitrogen atmosphere.

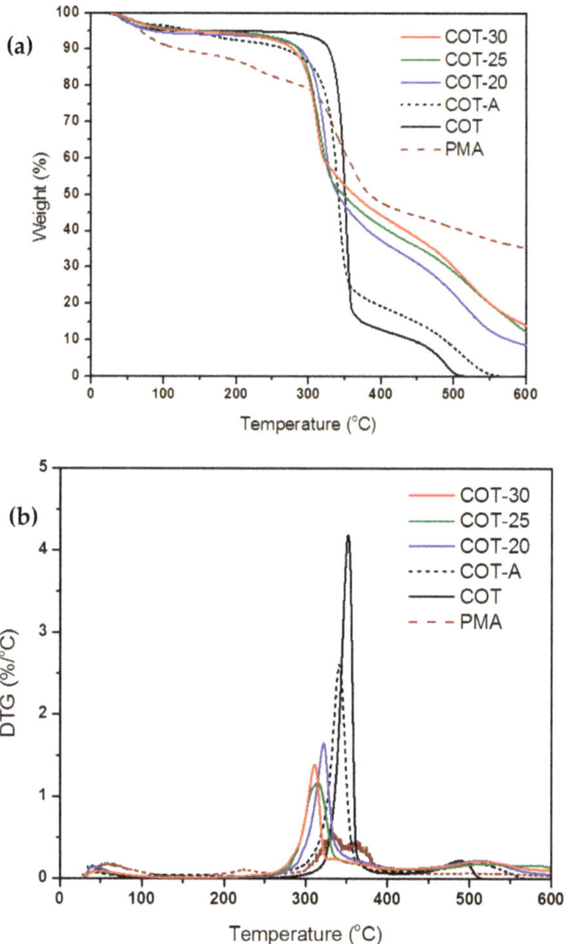

Figure 8. TGA (**a**) and DTG (**b**) curves of the neat cotton and treated cotton in an oxygen atmosphere.

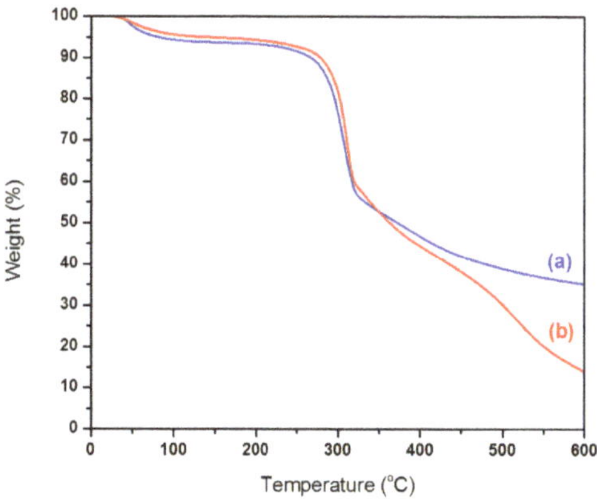

Figure 9. TGA curves of COT-30 in a nitrogen (**a**) and oxygen (**b**) atmosphere.

To fortify for the above assertion, the charred fabrics were characterized by ATR-FTIR, SEM, and EDX analyses. Comparison of the ATR-FTIR spectra of COT-30 (a) and COT-30-char at 600 °C (b) (Figure 10) revealed that the formation of unsaturated carbon–carbon bonds and the phosphorus containing charred residues occurred at elevated temperatures. The disappearance of the vibration at 2950 cm^{-1} during heating confirmed the loss of –CH$_2$– groups. The formation of a new peak at around 1689 and 1575 cm^{-1}, corresponding to the vibration modes of C=C bonds in the aromatic group, indicated the formation of graphite [26]. Moreover, the significant peak at 1193 cm^{-1} (medium) was assigned to the vibration of P=O bonds in the phosphate moieties. Therefore, we concluded from these findings that phosphorylated cellulose occurred under thermal degradation to promote thermally stable residue formation. Similar findings were concluded by Gaan et al. [27] for the actions of various phosphonate derivatives on cotton fabrics.

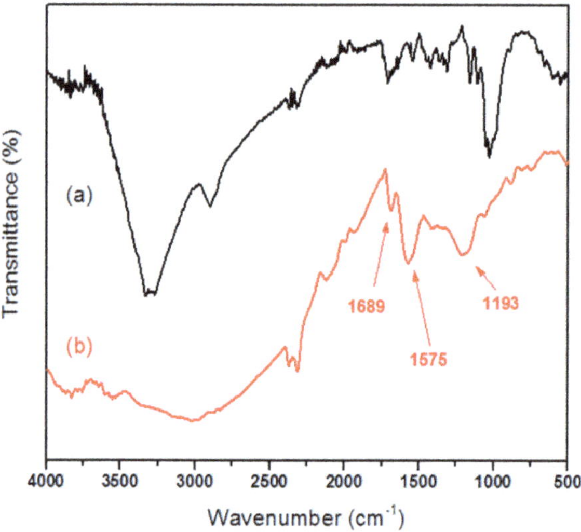

Figure 10. ATR-FTIR spectra of COT-30 (**a**) and COT-30-char (**b**).

We also consolidated other evidence from the EDX analysis and SEM images of char residues of COT-20, COT-25, and COT-30. SEM images of the surface morphology of char residues (Figure 11) displayed the morphology of the cotton coated with different FRs loadings. Extremely thin char fibers were obtained for all samples. The pieces of cotton covered with lower FR loadings tended to produce fiber clusters and became fiber chars (Figure 11A). The fiber clusters moved closer, and the fiber char became bigger with higher FR loadings (Figure 11B). The addition of FR to cotton fabrics resulted in increased fiber chars. The most abundant fibers arranged close together were found in the char fibers of COT-30 (Figure 11C). Figure 11A2,B2,C2 show that the size of the char fibers increased when the loadings of the FR coated on the cotton increased.

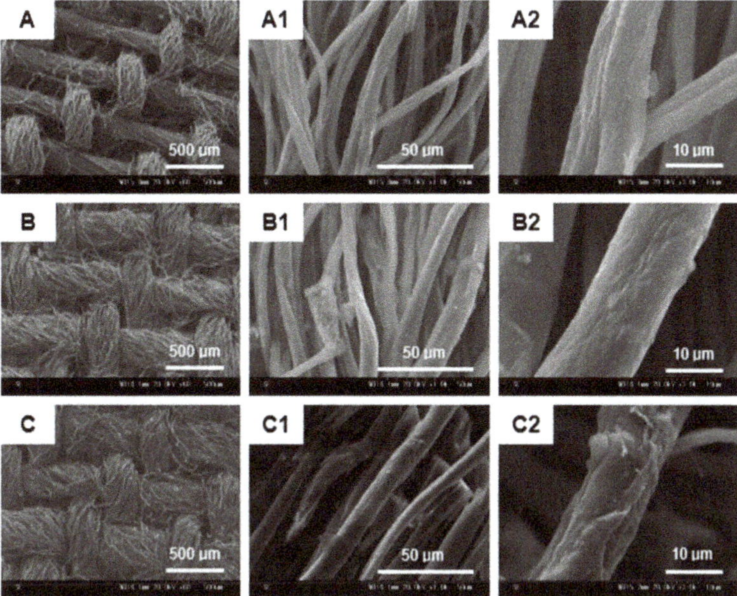

Figure 11. SEM images of char residues of COT-20 (**A**,**A1**,**A2**); COT-25 (**B**,**B1**,**B2**); and COT-30 (**C**,**C1**,**C2**).

Moreover, all EDX results presented in Figure 12 show that the char residues had very high carbon contents. We believed that the structure of the char fiber was mainly carbon backbone retained by the FRs. Additionally, the P contents in the treated fabrics before and after burning were also considered (Figure 13). We found that the P contents in the char residues were higher than the unburned fabric. Furthermore, the distribution of P elements became denser, and the density of P elements in the char residue increased dramatically with the increase in FR loadings (Figure 14) compared to the cotton fabrics without heating treatment.

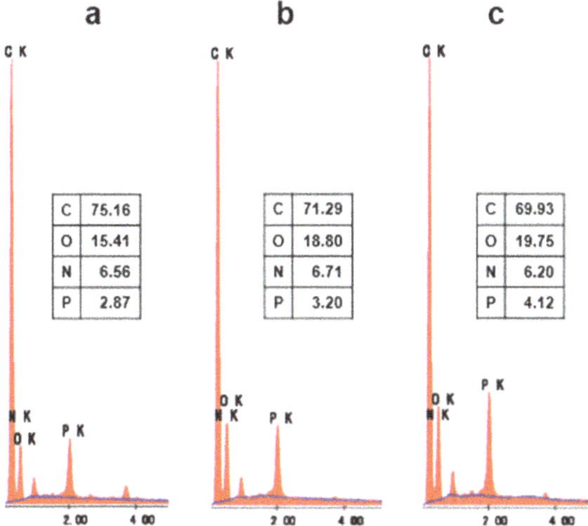

Figure 12. Energy-dispersive X-ray spectroscopy (EDX) results of the char residue of COT-20 (**a**), COT-25 (**b**), and COT-30 (**c**).

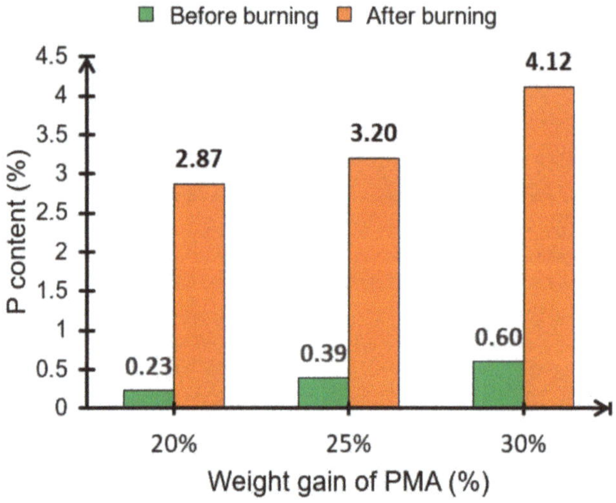

Figure 13. P content in the treated fabrics before and after burning.

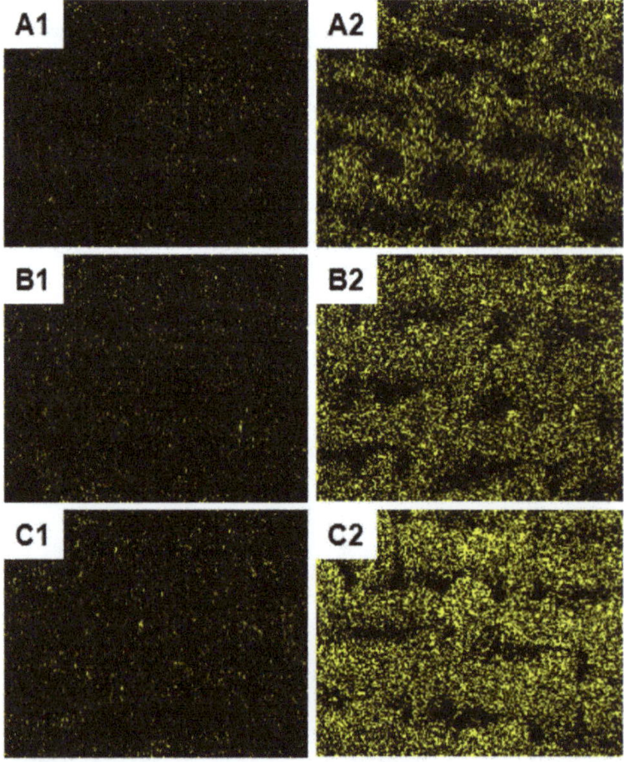

Figure 14. Mapping of the P element in char residue of COT-20 (**A1,A2**); COT-25 (**B1,B2**); COT-30 (**C1,C2**) before (left) and after (right) burning.

4. Conclusions

A novel FR formulation was synthesized successfully using a simple process. The precise structure of FRs also was elucidated by FTIR and ^{1}H-NMR spectroscopy analyses. The results of the ATR-FTIR, SEM, and EDX study showed that the PMA system monomer was coated onto the surface of the cotton fabrics. FRs promoted thermally stable char formation. The thick char fibers containing a uniform distribution of P elements were obtained from the FR coated cotton samples. The thermal stability of the PMA coated cotton fabrics was significantly enhanced. The PMA system used as a durable flame retardant exhibited high effectiveness of the condensed phase for cotton fabrics. We observed that the system passed the DIN 53906 standard with a 25% add-on.

Author Contributions: Conceptualization, H.K.N. and C.N.; Investigation, H.K.N., Analysis, H.K.N. and W.S.; Writing-Original Draft Preparation, H.K.N; Writing-Review & Editing, all authors. All authors have read and agreed to the published version of the manuscript.

Funding: This research was financially supported by the VNU-HCMC Scientific Research Foundation (Grant number **C2016−18−19**).

Acknowledgments: The authors would like to acknowledge the JSPS Core-to-Core Program, Kyoto Institute of Technology, Japan for their assistance with the measurement techniques.

Conflicts of Interest: The authors have declared that no competing interests exist.

References

1. Wakelyn, P.J.; Bertoniere, N.R.; French, A.D.; Thibodeaux, D.P.; Triplett, B.A.; Rousselle, M.; Goynes, W.R.; Edwards, J.V.; Hunter, L.; McAlister, D.D.; et al. *Cotton Fiber Chemistry and Technology*; CRC Press: Boca Raton, FL, USA, 2006.
2. Gaan, S.; Salimova, V.; Rupper, P.; Ritter, A.; Schmid, H. *Flame Retardant Functional Textiles*; Woodhead Publishing Limited: Cambridge, UK, 2011.
3. Xing, W.; Jie, G.; Song, L.; Hu, S.; Lv, X.; Wang, X.; Hu, Y. Flame retardancy and thermal degradation of cotton textiles based on UV-curable flame retardant coatings. *Thermochim. Acta* **2011**, *513*, 75–82. [CrossRef]
4. Placido, J.; Capareda, S. Production of silicon compounds and fulvic acids from cotton wastes biochar using chemical depolymerization. *Ind. Crop. Prod.* **2015**, *67*, 270–280. [CrossRef]
5. Przybylak, M.; Maciejewski, H.; Dutkiewicz, A.; Wesolek, D.; Przybylak, M.W. Multifunctional, strongly hydrophobic and flame-retarded cotton fabrics modified with flame retardant agents and silicon compounds. *Polym. Degrad. Stab.* **2016**, *128*, 55–64. [CrossRef]
6. Giraud, S.; Bourbigot, S.; Rochery, M.; Vroman, I.; Tighzert, L.; Delobed, R. Flame Behavior of Cotton Coated with Polyurethane Containing Microencapsulated Flame Retardant Agent. *J. Ind. Text.* **2001**, *31*, 11–26. [CrossRef]
7. Liu, W.; Chen, L.; Wang, Y.Z. A novel phosphorus-containing flame retardant for the formaldehyde-free treatment of cotton fabrics. *Polym. Degrad. Stab.* **2012**, *97*, 2487–2491. [CrossRef]
8. Liang, S.; Matthias, N.; Neisius, N.S.; Gaan, S. Recent developments in flame retardant polymeric coatings. *Progr. Org. Coat.* **2013**, *76*, 1642–1665. [CrossRef]
9. Weil, E.D.; Levchik, S.V. *Flame Retardants for Plastics and Textiles*; Carl Hanser Verlag GmbH Co KG.: Munich, Germany, 2015.
10. Nguyen, C.T.; Kim, J.W. Thermal stabilities and flame retardancies of nitrogen-phosphorus flame retardants based on bisphosphoramidate. *Polym. Degrad. Stab.* **2008**, *93*, 1037–1043. [CrossRef]
11. Nguyen, C.T.; Kim, J.W. Synthesis of a Novel Nitrogen-Phosphorus Flame Retardant Based on Phosphoramidate and Its Application to PC, PBT, EVA, and ABS. *Macromol. Res.* **2008**, *16*, 620–625. [CrossRef]
12. Hai, V.T.; Nguyen, C.T.; Lee, K.H.; Kim, J.W. Thermal stability and flame retardancy of novel phloroglucinol based organo phosphorus compound. *Polym. Degrad. Stab.* **2010**, *95*, 1092–1098.
13. Yuan, H.; Xing, W.; Zhang, P.; Song, L.; Hu, Y. Functionalization of Cotton with UV-Cured Flame Retardant Coatings. *Ind. Eng. Chem. Res* **2012**, *51*, 5394–5401. [CrossRef]

14. Edwards, B.; El-Shafei, A.; Hauser, P.; Malshe, P. Towards flame retardant cotton fabrics by atmospheric pressure plasma-induced graft polymerization: Synthesis and application of novel phosphoramidate monomers. *Surf. Coat. Technol.* **2012**, *209*, 73–79. [CrossRef]

15. Cheema, H.A.; El-Shafei, A.; Hauser, P.J. Conferring flame retardancy on cotton using novel halogen-free flame retardant bifunctional monomers: Synthesis, characterizations and applications. *Carbohydr. Polym.* **2013**, *92*, 885–893. [CrossRef] [PubMed]

16. Wang, S.; Sui, X.; Li, Y.; Li, J.; Xu, H.; Zhong, Y.; Zhang, L.; Mao, Z. Durable flame retardant finishing of cotton fabrics with organosilic on functionalized cyclotriphosphazene. *Polym. Degrad. Stab.* **2016**, *128*, 22–28. [CrossRef]

17. Chen, Q.; Yang, C.Q.; Zhao, T. Heat release properties and flammability of the nylon/cotton blend fabric treated with a crosslinkable organophosphorus flame retardant system. *J. Anal. Appl. Pyrolysis* **2014**, *110*, 205–212. [CrossRef]

18. Li, X.H.; Chen, H.Y.; Wang, W.T.; Liu, Y.Q.; Zhao, P.H. Synthesis of a formaldehyde-free phosphorus–nitrogen flame retardant with multiple reactive groups and its application in cotton fabrics. *Polym. Degrad. Stab.* **2015**, *120*, 193–202. [CrossRef]

19. Dong, C.H.; Lu, Z.; Wang, P.; Zhu, P.; Li, X.C.; Sui, S.Y.; Zhang, L.; Liu, J. Flammability and thermal properties of cotton fabrics modified with a novel flame retardant containing triazine and phosphorus components. *Text. Res. J.* **2017**, *87*, 1367–1376. [CrossRef]

20. Salmeia, K.A.; Gaan, S.; Malucelli, G. Recent Advances for Flame Retardancy of Textiles Based on Phosphorus Chemistry. *Polymers* **2016**, *8*, 319. [CrossRef]

21. Ivanov, B.E.; Karpova, T.I. Synthesis and properties of hydroxymethyl phosphonic and bishydroxymethylphosphinic acids. *Russ. Chem. Bull.* **1964**, *13*, 1140–1142. [CrossRef]

22. *Testing of Textiles-Determination of the Burning Behaviour-Vertically Method, Ignition by Application of Flame to Base of Specimen*; DIN 53906:1974-02; Beuth Verlag: Berlin, Germany, 1974.

23. *Textiles-Tests for Colour Fastness-Part C10: Colour Fastness to Washing with Soap or Soda and Soda*; IS/ISO 105-C10:2006; International Organization for Standardization: Geneva, Switzerland, 2006.

24. Jimenez, M.; Guin, T.; Bellayer, S.; Dupretz, R.; Bourbigot, S.; Grunlan, J.C. Microintumescent mechanism of flame-retardant water-based chitosan–ammonium polyphosphate multilayer nanocoating on cotton fabric. *J. Appl. Polym. Sci.* **2016**, *133*, 43783. [CrossRef]

25. Price, D.; Horrocks, A.R.; Akalin, M.; Faroq, A.A. Influence of flame retardants on the mechanism of pyrolysis of cotton (cellulose) fabrics in air. *J. Anal. Appl. Pyrolysis* **1997**, *40*, 511–524. [CrossRef]

26. Qi, X.; Qu, J.; Zhang, H.B.; Yang, D.Z.; Yu, Y.H.; Chi, C.; Yu, Z.Z. FeCl Intercalated Few-Layer Graphene for High Lithium-Ion Storage Performance. *J. Mater. Chem. A* **2015**, *3*, 15489–15504. [CrossRef]

27. Gaan, S.; Rupper, P.; Salimova, V.; Heuberger, M.; Rabe, S.; Vogel, F. Thermal decomposition and burning behavior of cellulose treated with ethyl ester phosphoramidates: Effect of alkyl substituent on nitrogen atom. *Polym. Degrad. Stab.* **2009**, *94*, 1125–1134. [CrossRef]

Article

Thermal Stability and Flammability Behavior of Poly(3-hydroxybutyrate) (PHB) Based Composites

Henri Vahabi [1,2,*], Laurent Michely [3], Ghane Moradkhani [1,2], Vahideh Akbari [1,2], Marianne Cochez [1,2], Christelle Vagner [1,2,4], Estelle Renard [3], Mohammad Reza Saeb [1,2] and Valérie Langlois [3,*]

[1] Université de Lorraine, CentraleSupélec, LMOPS, F-57000 Metz, France
[2] Laboratoire Matériaux Optiques, Photoniques et Systèmes, CentraleSupélec, Université Paris-Saclay, 57070 Metz, France
[3] Université Paris Est, Institut de Chimie et des Matériaux Paris-Est, UMR 7182 CNRS-UPEC, 2, rue Henri Dunant, 94320 Thiais, France
[4] Aix Marseille University, CNRS, MADIREL UMR 7246, F-13397 Marseille, France
[*] Correspondence: henri.vahabi@univ-lorraine.fr (H.V.); langlois@icmpe.cnrs.fr (V.L.); Tel.: +33-38-793-9186 (H.V.); +33-14-978-1217 (V.L.); Fax: +33-38-793-9101 (H.V.); +33-14-978-1217 (V.L.)

Received: 23 May 2019; Accepted: 10 July 2019; Published: 11 July 2019

Abstract: A series of samples based on poly(3-hydroxybutyrate) (PHB) containing five different additives were prepared and their thermal stability and flammability were discussed. The samples first underwent flammability screening by using Pyrolysis Combustion Flow Calorimeter (PCFC) analyses. Then, four samples were selected for further investigations. PHB composites containing sepiolite (Sep.) inorganic nanofiller, and also organic ammonium polyphosphate (APP) were examined for flammability and thermal behavior using PCFC, thermogravimetric analysis (TGA), flame test, and Differential Scanning Calorimetry (DSC) analyses. Moreover, burning behavior of samples were captured on a digital camera to give a deeper sense of their flammability character for comparison. The results revealed a significant improvement of flammability and thermal stability of composites, particularly in the presence of sepiolite with respect to the value obtained for unfilled PHB. Regarding TGA results, the char residue yield was increased to ca. 20.0 wt.% in the presence of sepiolite, while 0.0 wt.% was observed for PHB. PCFC measurements uncovered higher performance of PHB-Sep. sample as signaled by 40% reduction in the peak of heat release rate with respect to PHB. According to observations, PHB-Sep. sample showed non-dripping behavior with high capacity of charring in the presence of Sep. in a vertical flame test.

Keywords: poly(3-hydroxybutyrate) (PHB); flame retardancy; microcalorimetry of combustion

---

1. Introduction

Biodegradable biopolymer is a general term used for polymers that are synthesized from natural resources and can be degraded/decomposed by micro-organisms -what positioned them in front of fossil-based polymers in the frame of attention from an environmental perspective [1,2]. A recent survey revealed huge growth in the number of publications on bio-based polymers [3]. In polymer science and technology, two generations of biodegradable polyesters have been identified: poly(lactic acid) (PLA) and poly(hydroxyl alkanoate) (PHAs) [4]. Though PLA has already celebrated its maturity age, detailed analyses foresee a flourishing future of PHA in the global plastics market [4]. PHAs have exceptional characteristics such as optical activity, biocompatibility, a barrierity which is higher than PLA, and full biodegradability [5]. Moreover, synthesis of PHAs from bacteria eliminates the need for fermentation of food resources required in PLA production [6,7]. Furthermore, PHA is fully biodegradable in soil and marine water as well as in home compost. Researchers and engineers alike identified benefits of PHAs to keep it in competition with PLA thanks to the microbial features of PHAs [8]. On the other hand, high production cost, low thermal resistance, high flammability, poor mechanical properties compared to conventional polymers and the limited functionality of PHAs are the main reasons behind accelerative growth in research on PHAs [9–11]. Although the main application of PHAs is in the biomedical sector, a future prospect for PHAs can be imagined due to two of its exceptional characteristics: full biodegradability and a synthesis ability from bacteria [4]. In this sense, improvement of flammability properties of PHAs seems to be a priority for some applications [12]. Poly(3-hydroxybutyrate) (PHB) as the first member of PHAs family has been the candidate of many research programs [13]. Improvement of flame retardancy of some copolymers of PHB has been already studied by the incorporation of renewable raw materials, [14], melamine phosphate modified lignin [15], layered double hydroxides [16], kenaf fiber [17,18], halloysite nanotubes [19] aluminum phosphinate in combination with nanometric iron oxide and antimony oxide [10]. However, to the best of our knowledge, flame retardancy of PHB solely was not the subject of reports.

In this work, PHB and its various composites containing several additives were prepared and their behavior in terms of thermal decomposition and flammability was investigated. Following a preliminary work including twelve formulations, a series of composites containing sepiolite as well-known typical of mineral nanofillers, organic ammonium polyphosphate (APP) as phosphorus additive, and several natural additives including lignin and starch were developed individually and combinatorial. The amount of additive was kept constant while varying ingredient compositions in a PHB matrix to comparatively evaluate their flame scenario. It is well-known that sepiolite is a needle-like-shaped nanoclay and is biocompatible [20]. It has been already used in PHB and its copolymers in order to improve thermal and/or mechanical properties [21,22]. It was also proved that sepiolite as an inorganic biocompatible mineral can reveal promising features in combination with APP conventional flame retardant [23–25]. The combined use of the aforementioned additives from organic and inorganic families with micron and nano-size scales together with lignin and starch enabled us to understand the flammability behavior of PHB in different situations and provided a basis for understanding the degree to which PHB can be armed with additives to resist against fire. The prepared composites were subjected to a variety of characterizations. Pyrolysis combustion flow calorimeter (PCFC) was used as the first estimation technique to screen samples in terms of flammability. Moreover, thermogravimetric analysis (TGA), the direct flame test, scanning electron microscopy (SEM), size-exclusion chromatography (SEC), differential scanning calorimetry (DSC), and X-ray diffraction (XRD) measurements were performed.

2. Materials and Methods

Poly(3-hydroxybutyrate) (PHB) powder was supplied by BIOMER (PHB T19, Krailling Germany). Ammonium polyphosphate (APP) was purchased from Clariant (Muttenz, Switzerland, Exolit AP 423, $[NH_4PO_3]_n$, n > 1000, particle size $\approx$ 8 μm, specific surface area 1.1 m^2/g). Sepiolite ($Mg_4Si_6O_{15}(OH)_2.6H_2O$), namely Sep., was provided by Tolsa, Spain, (Pangel S9), and used without any modification. Starch (soluble, ACROS Organics™) was provided by Fisher Scientific (Hampton, NH, USA). Moreover, lignin (alkali) was purchased from Sigma-Aldrich (CAS Number: 8068-05-1).

PHB alone and together with additives was melt mixed in a Xplore conical twin screw microcompounder (MC 15 HT, Netherlands) at 175 °C and a rotor speed of 80 rpm during 5 min. First, PHB was fed into the mixing chamber and then after melting, additives were added. After the melt mixing was completed, the samples were casted as a thin film by using casting machine (Xplore-thickness: 50 μm, width: 3 cm). The sample names and compositions are given in Table 1. Loading percentage of additives was fixed at 15 wt.%. As mentioned earlier, sepiolite as a natural classic fire retardant in combination with APP and natural polymers was examined for flammability with PCFC first estimations. Lignin and starch were chosen as bio-based sources of carbon for flame retardant systems. These materials have already been used in PHB [26–29], however their combination with APP has never been reported.

Table 1. Names and compositions of the PHB-based composite samples prepared in this study.

Number	Sample Code	PHB	Sepiolite	APP	Lignin	Starch
1	PHB	100	0	0	0	0
2	PHB-Sep.	85	15	0	0	0
3	PHB-APP	85	0	15	0	0
4	PHB-Lig.	85	0	0	15	0
5	PHB-Starch	85	0	0	0	15
6	PHB-APP-Lig.	85	0	10	5	0
7	PHB-APP-Sep.	85	5	10	0	0
8	PHB-APP-Starch	85	0	10	0	5
9	PHB-APP-Lig.-Sep.	85	2.5	10	2.5	0
10	PHB-APP-Lig.-Starch	85	0	10	2.5	2.5
11	PHB-APP-Sep.-Starch	85	2.5	10	0	2.5

The cross-section of selected samples was analyzed in a scanning electron microscope (SEM) manufactured by Carl Zeiss (Oberkochen, Germany) with a Field Emission Gun (FEG) at a magnification between 1.00K× et 5.00K× and at a working distance of 9–10 mm. The signal of the secondary electron was collected to scanning the sample. The voltage was 5 kV with a probe from 300 pA. The cross-sections were prepared by manual fracture in the transversal direction. Subsequently, the cross sections were metalized by a sputtering process with platinum (3.0 nm). The samples were also analyzed using an EDX (Energy-dispersive X-ray spectroscopy) microanalysis system (X-ray photon dispersion analysis). The EDX microanalysis system is an advanced Aztec EDS system, provided by Oxford Instruments (Abingdon-on-Thames, UK). The X detector is a 50 mm^2 X-max SDD detector. Structural characterization was performed by X-ray diffraction (XRD) using a D8 advance Bruker diffractometer (Cu Kα radiation). Data were recorded over a 2θ range from 10 to 30° by step of 0.0102° at an incident wavelength λ of 1.54056 Å. From the patterns, the cell parameters a, b, and c of the orthorhombic unit cell were calculated from the maxima of the (040), (110), and (121) peaks according to Equation (1):

$$\frac{1}{d^2} = \frac{h}{a^2} + \frac{k}{b^2} + \frac{l}{c^2} \tag{1}$$

where h, k, and l are the Miller crystallographic indexes and d is the interplanar spacing defined by Bragg's law d = λ/(2sinθ) with θ the scattering angle and λ the wavelength of the incident wave (=1.54056 Å). From these values, the lattice volume V was calculated according to the volume of an orthorhombic unit cell: V = a × b × c.

Thermal characterization was performed using Perkin Elmer Diamond Differential scanning calorimetry (DSC) with the following procedure: a first heating run from −60 to 200 °C with a heating rate of 20 °C/min was performed to determine the melting temperature (T_m) and the melting enthalpy (ΔH_m), followed by a cooling run to −60 °C with a cooling ramp of 200 °C/min. Then a second heating run from −60 to 200 °C at 20 °C/min was performed. The glass transition temperature (T_g) was obtained in the second heating. The degree of crystallinity χ_c was calculated as a function of the real amount of PHB according to Equation (2):

$$\chi_c = \frac{1}{w_{PHB}} \times \frac{\Delta H_m}{\Delta H^{\circ}_m} \times 100 \tag{2}$$

where ΔH_m is the specific enthalpy of melting of the sample studied, w_{PHB} is the weight fraction of the PHB in the blend, and ΔH°_m represents the specific enthalpy of melting for the 100% crystalline PHB, taken as 146 J/g.

Polymer molar masses were determined by Size-exclusion chromatography (SEC) using a Kontron 420 pump with 2 styragel columns connected in series which type is PL gel (mixte C) from polymer laboratories, and a Shodex RI-71 model refractive index detector (Japan). $CHCl_3$ was used as eluent at a flow rate of 1.0 mL/min. A calibration curve was generated with polystyrene standards of low polydispersity purchased from Polysciences (Germany). Thermogravimetric analysis (TGA) was performed using a Setaram Labsys Evo thermogravimetric analyzer (France), under nitrogen with a heating rate of 10 °C/min.

Flammability properties were investigated using Pyrolysis Combustion Flow Calorimeter (PCFC) instrument (FTT Company, UK). The samples (1 to 4 mg) were heated at 1 °C/s from 20 °C to 750 °C in a pyrolyzer and the degradation products were conducted to another chamber and mixed with oxygen. Combustion took place at 900 °C. Each sample was tested 3 times and the related accuracy was around 5%. Vertical burning test (unnormalized) was also carried out on prepared film samples. Samples dimension was 120 mm × 25 mm × 50 μm. Samples were placed vertically inside a frame and exposed to a Bunsen burner flame for 3 s. A descriptive scheme of test is presented in Section 3.8. Moreover, digital video of tests was recorded and the selected images were also extracted. All videos are available in the Supplementary Materials.

3. Results

3.1. Premilinary Test by PCFC: Screening for Flammability Estimation

A series of formulations containing APP, sepiolite, starch, lignin, and their combinations were prepared, Table 1. The goal of this preliminary work was to screening various formulations in terms of flammability and then selection of promising samples for the second-step investigation. Pyrolysis Combustion Flow Calorimeter (PCFC) is well known as a useful apparatus for screening the flammability of materials [30–33]. PHB and its composites defined in Table 1 were analyzed using PCFC. The obtained curves, heat of release rate (HRR) as a function of temperature, are presented in Figure 1. The presence of lignin and starch significantly decreased the temperature of peak of heat release (pHRR) rate and the reduction in pHRR was not significant in the presence of these additives in respect to that of PHB. The decrease in temperature of pHRR (T_{pHRR}) can be attributed to degradation of PHB in presence of these fillers [28]. In the case of sepiolite, there is less difference between T_{pHRR} for PHB-Sep. and PHB. Moreover, the reduction of pHRR is more important. APP acts essentially in the condensed phase. The incorporation of APP was led to decrease in pHRR, close to the performance of sepiolite. However, it had no serious effect on T_{pHRR} compared to PHB. The combination of APP with sepiolite, lignin and starch increased T_{pHRR}, while the value of pHRR was approximately near to that of other composites. Since the performance features of lignin and starch were less abundant than sepiolite, it was decided to conduct further investigation solely on PHB, PHB-Sep., PHB-APP, and PHB-APP-Sep. samples, Table 2.

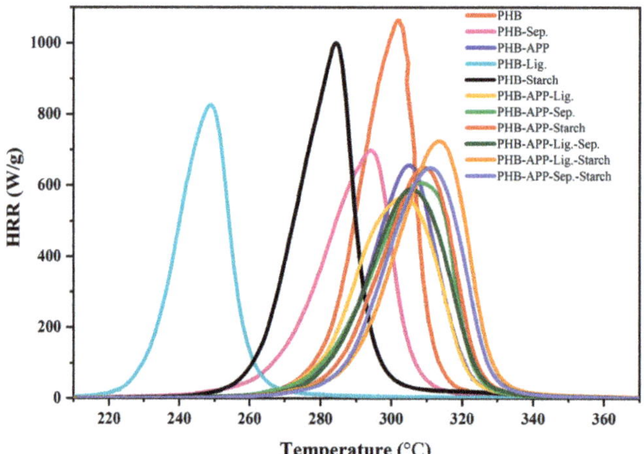

Figure 1. Heat Release Rate (HRR) curves obtained in PCFC tests.

Table 2. Names and compositions of selected samples after screening study.

Sample Code	PHB	Sepiolite	APP
PHB	100	0	0
PHB-APP	85	0	15
PHB-Sep.	85	15	0
PHB-APP-Sep.	85	5	10

3.2. Morphology Investigations (SEM)

Figure 2 displays the SEM images of PHB and its composites. The micrograph of PHB shows a homogenous structure with only few structural defects, Figure 2a. Sepiolite nanoparticles were homogeneously dispersed in form of fibrous particle in PHB, Figure 2b. The size of APP particles was approximatively between 1 and 10 µm, Figure 2c. The adhesion of APP particles to the matrix seems not to be optimized. However, a better adhesion can be observed between APP and matrix in the case of a PHB-APP-Sep. sample, Figure 2d. Sepiolite particles are also homogeneously dispersed in nanometric size.

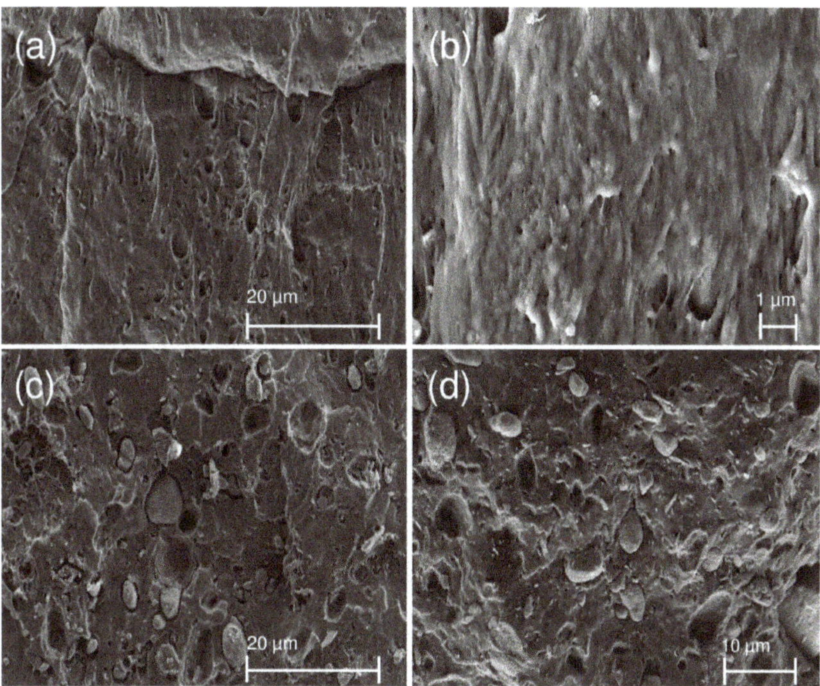

Figure 2. SEM images of PHB and its composites: (**a**) PHB; (**b**) PHB-Sep.; (**c**) PHB-APP; and (**d**) PHB-APP-Sep.

EDX analyses performed on the PHB-APP-Sep. blend determined the elements characterizing sepiolite and APP (Figure 3). Based on SEM/EDX analysis in different areas, it is possible to better define the dispersion of sepiolite and APP (Table 3). In zone 1, the presence of peaks C and O confirms that the phase is mainly composed of PHB. Within this matrix, fine sepiolite needles are dispersed in zone 2 where the Si and Mg elements detected by EDX analysis show the presence of sepiolite. In zone 3, the presence of the elements P, Si and Mg attest the presence of both APP and sepiolite. There would therefore be an aggregation of the two types of charge with PHB with a majority of APP.

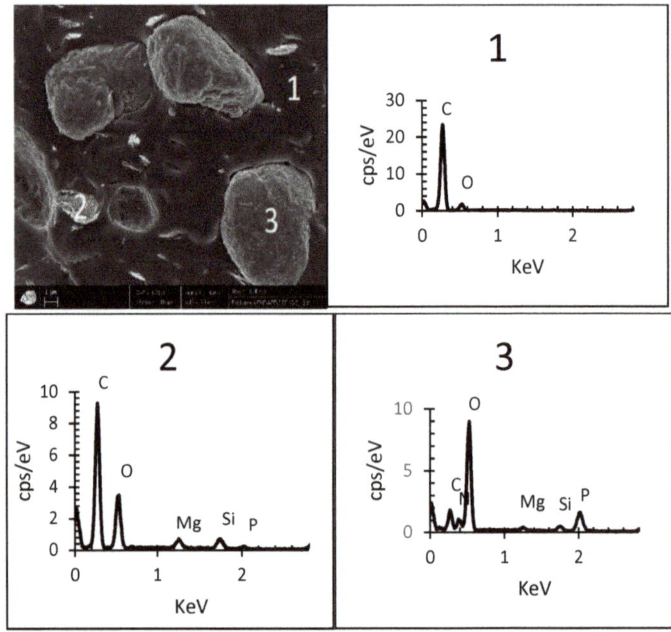

Figure 3. SEM images of PHB- APP-Sep. sample and analyzed zones by EDX.

Table 3. EDX analysis of PHB-APP-Sep. sample on different areas.

Area Number	Elements (%wt.) Normalized at 100					
	C	O	Si	Mg	P	N
Area 1	90.7	9.0	0.3	-	-	-
Area 2	66.4	24.9	4.6	2.5	1.6	-
Area 3	14.4	54.8	2.7	1.1	15.0	12.0

3.3. Differential Scanning Calorimetry (DSC)

Differential scanning calorimetry (DSC) analysis was carried out on all samples and the obtained results recorded during the first and second heating cycles are presented in Figure 4 and Table 4. The glass transition temperature (Tg), the crystallization temperature (Tcc) and the melting temperature (T_m) of PHB were found to be 5.3, 48.3 and 172.6 °C, respectively. These characteristic temperatures were not greatly modified by the incorporation of additives indicating that the presence of fillers does not affect the global morphology of PHB. However, the addition of additives, especially for sepiolite, led to a decrease in the crystallinity of PHB from 58 to 48 °C and probably led to a change in the size of the crystalline zones. The PHB-Sep blend display double melting peaks at 171.9 and 175 °C compared to pure PHB with one single melting peak at 172.2 °C. This is while T_m were affected more obviously due to sensitivity of PHB melting to additive addition. The presence of double endothermic melting peaks in the PHB composites was ascribed to the melt recrystallization mechanism [34].

All thermograms showed a crystallization peak between 45 °C and 48 °C during the second heating. This peak occurring after glass transition (T_g) is due to a resumption of mobility of the molecules allowing the end of unfinished crystallization of the polymer during rapid cooling. In the presence of APP and sepiolite, the ΔH_{Tcc} of PHB decreased. When using both APP and sepiolite, this peak lost its amplitude (signaled by a decrease in ΔH_{Tcc}), compared to the peak appearing for PHB. Such a behavior was completely changed by addition of APP to the formulation due to the pace formation of PHB crystals formed during the cooling process. All in all, the presence of additives did not affect Tg and Tm of PHB. Moreover, the decrease of crystallization enthalpy during the second heating rate was also inferred on account of the presence of additives and formation of crystalline domains during the cooling process after the first heating step.

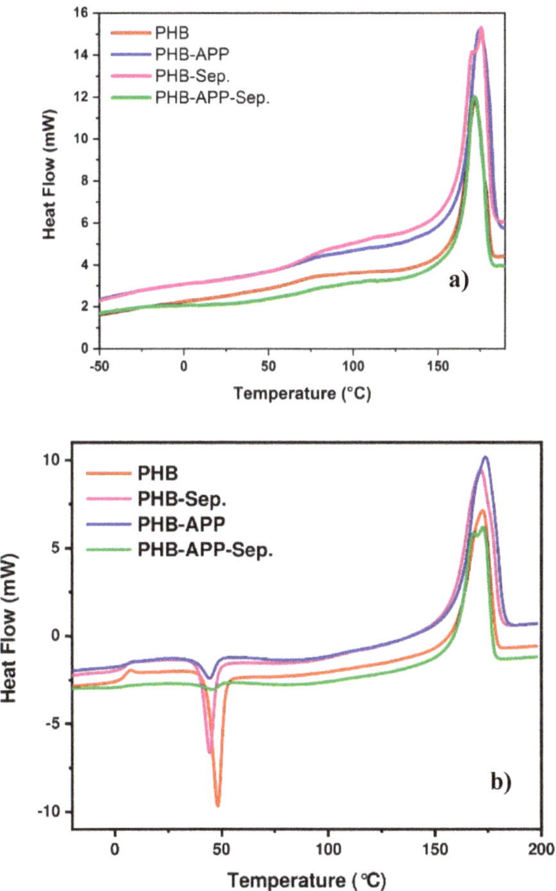

Figure 4. DSC curves for PHB, PHB-Sep., PHB-APP, PHB-APP-Sep. samples, (**a**) 1st heating (**b**) 2nd heating.

Table 4. DSC curves data of PHB and all composites (The values of enthalpy of melting for blends are normalized in compositions).

Sample	1st Heating			2nd Heating			
	T_{m1} (°C)	ΔH_{Tm1} (J/g)	X_c (%)	T_g (°C)	T_{cc} (°C)	ΔH_{Tcc} (J/g)	T_{m2} (°C)
PHB	172.2	85.4	58.2	5.30	48.3	36.7	172.6
PHB-APP	174.9	84.6	57.7	6.50	44.6	5.0	173.9
PHB-Sep.	171.9	70.8	48.3	5.70	44.6	14.2	171.5
PHB-APP-Sep.	175.9	73.3	50.0	4.15	45.9	2.4	172.9

3.4. X-Ray Diffraction

The crystal structure of PHB and the blends were analyzed by the Wide-angle X-ray scattering (WAXD) of α-form crystals, Figure 5. The two peaks, at about 15° and 16° for PHB-APP and PHB-APP-Sep.

are diffraction patterns corresponding to APP rather than PHB crystals. It is obvious that the diffraction patterns and diffraction angles of the four specimens are almost similar (Table 5). The lattice parameters calculated by using interplanar spacing of planes (040), (110), and (121) in WAXD are shown in Table 6. The lattice parameters for PHB-Sep. are slightly different from those of PHB but the lattice for PHB-APP and PHB-APP-Sep. are similar to those of PHB. Particles of sepiolite slightly affected the basic crystal lattice. These results were supported by the lower values of melting enthalpy in the presence of sepiolite, which was determined by DSC.

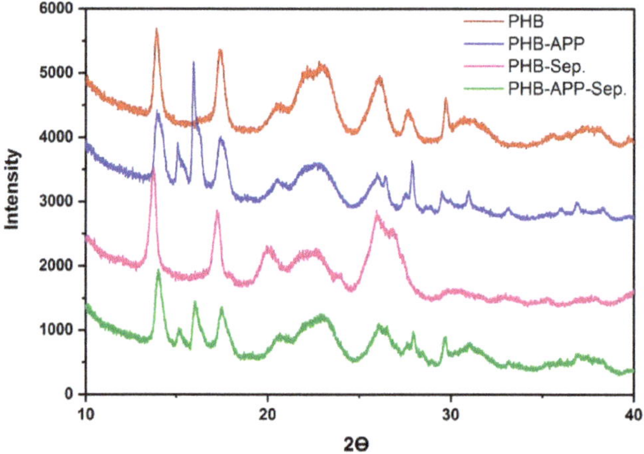

Figure 5. XRD patterns obtained from PHB; PHB-APP; PHB-Sep.; PHB-APP-Sep.

Table 5. XRD peaks of PHB and blends.

Sample Code	2θ (°)					
	(020)	(110)	(021)	(111)	(121)	(040)
PHB	13.91	17.41	20.46	23.06	26.06	27.66
PHB-APP	14.02	17.41	20.41	22.94	25.98	27.63
PHB-Sep.	13.75	17.26	19.93	22.66	26.07	27.04
PHB-APP-Sep.	14.04	17.42	20.54	23.04	26.05	27.83

Table 6. Lattice constants and lattice volume of PHB and blends.

Sample Code	Lattice Constants (Å) & Lattice Volume (Å)			
	a	b	c	v
PHB	5.54	12.88	5.87	4.19
PHB-APP	5.54	12.90	5.92	4.23
PHB-Sep.	5.54	13.17	5.78	4.20
PHB-APP-Sep.	5.54	12.81	5.90	4.19

3.5. Size-Exclusion Chromatography (SEC)

The impact of the incorporation of additives into the PHB was further studied by SEC. The obtained values in terms of number-average molecular weight (Mn), weight-average molecular weight (Mw), and polydispersity index (PDI) are given in Table 7. The M_n and M_w values were decreased in the presence of APP and sepiolite, either alone or combinatorial. However, the decrease in M_w was less important than M_n for all samples. Such a fall in Mn was more significant in the presence of sepiolite than APP. It was also possible that the blending procedure in the presence of additives contained small quantity of water that assisted in chain scission [6].

Table 7. Molecular weight of PHB and its corresponding blends.

Sample Code	M_n (g/mol)	M_w (g/mol)	PDI
PHB	106,000	254,400	2.4
PHB-APP	72,000	216,000	3.0
PHB-Sep.	55,000	148,500	2.7
PHB-APP-Sep.	69,000	186,300	2.7

3.6. Thermogravimetric Analysis (TGA)

Figure 6 displays TGA (a) and derivative thermogravimetric (DTG) (b) thermograms of the PHB and its composites under nitrogen atmosphere. The important parameters extracted from these curves are presented in Table 8. First, it is apparent from the curves that PHB and its composite samples reveal a one-stage decomposition behavior under nitrogen atmosphere. The onset temperature of thermal decomposition ($T_{10\%}$, temperature at which 10% weight loss takes place) was 264 °C for the PHB. Another phenomenon was recognized to take place around 300 °C, where PHB was completely degraded and no residual mass was remained. The main mechanism of thermal decomposition of PHB corresponds to β-elimination of PHB chains that facilitates the formation of crotonic acid, dimeric, trimeric and tetrameric volatiles [35–38]. The presence of additives did not substantially decreased the T_{onset}, in turn surprisingly led to a shift in decomposition temperature from 265 °C to 273 °C in the presence of APP and APP/Sep. Monitoring the maximum rate decomposition temperature from DTG curves (T_{max}) unraveled that T_{max} shifted to higher temperatures in the presence of APP and APP/Sep. Starting from 310 °C, the thermal stability of PHB-Sep. sample was significantly higher than other samples. This trend was observed until the end of TGA measurements. The char residue at 800 °C was 16 wt.% for PHB-Sep. sample, against 6 wt.% and 11 wt.% for PHB-APP and PHB-APP-Sep. samples, respectively. It is well known that the presence of a flame retardant, especially a phosphorus one, can affect onset the temperature of a polymer [39]. Surprisingly, sepiolite and APP did not decrease the T_{onset} of PHB. Even by combination of APP and sepiloite, T_{onset} was shifted to a higher temperature.

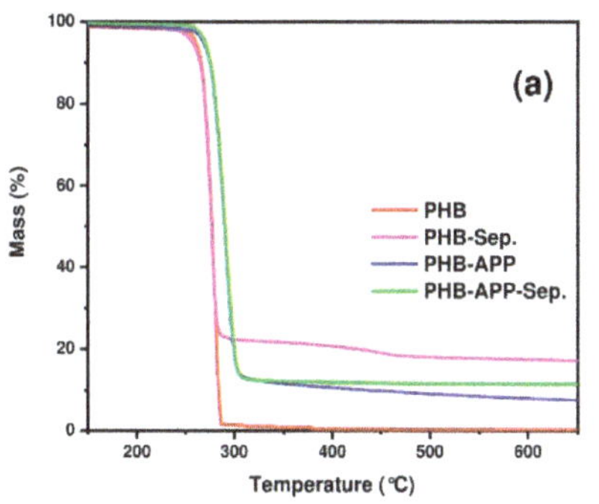
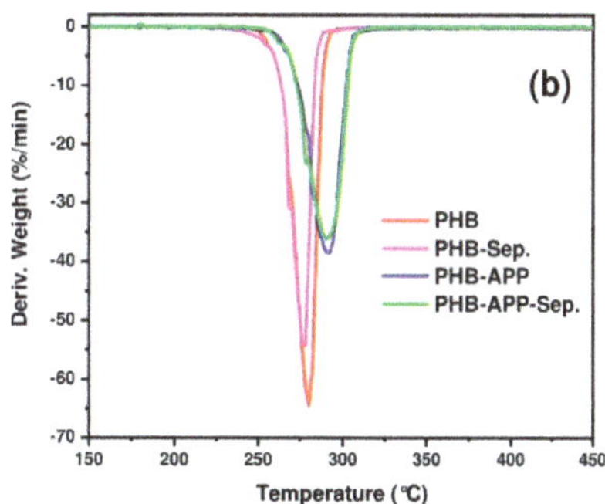

Figure 6. TGA thermograms of PHB and blends under nitrogen atmosphere. (**a**) TGA (**b**) DTG curves.

Table 8. TGA parameters PHB and blends in nitrogen (*maximum weight loss temperature, obtained from DTG).

Sample Code	$T_{10\%}$ (°C)	$T_{max}{}^*$ (°C)	Residue (wt.%)
PHB	265	280	0
PHB-Sep.	264	278	16
PHB-APP	274	292	6
PHB-APP-Sep.	273	290	11

3.7. Pyrolysis Combustion Flow Calorimeter (PCFC)

The flammability of PHB was studied and then compared with those of composites containing APP, Sep. and APP-Sep. using PCFC test. Figure 7 shows HRR curves of the studied samples, and the important data are extracted from these curves and summarized in Table 9. The temperature of pHRR was slightly increased in the presence of APP as well as when combination of APP and Sep. was used. Changes observed in PCFC curve of PHB upon introduction of APP, Sep., and APP/Sep. is significant, implying the sensitivity of flame retardancy of PHB to additives used in this work. A closer look at data reveals that the peak of HRR (pHRR) significantly dropped from 1064 to 656 W/g by addition of Sep. to PHB, showing a reduction of 38%. In the presence of APP, however, the pHRR reached the value 699 W/g (34% reduction compared to value for PHB). The combination of sepiolite and APP promisingly led to a significant drop in pHRR by 43%. The value of THR (total heat release) slightly decreased for composite samples compared to PHB. The lowest THR value was recognized for PHB-Sep. sample, taking value of 17 kJ/g. The use of APP increased THR value due to the contribution of APP function into gas phase. The temperature assigned to pHRR point for samples was almost between 302 and 308 °C, expect for PHB-Sep for which the temperature was 291 °C, which makes a similar sense to TGA measurements. It should be noticed that even with a peak of HRR at 600 kW/m^2, these samples can be considered as being flammable. However, the significant reduction in pHRR shows improvement in flammability behavior of these composites with respect to PHB. Although PCFC gives some indication about flammability of material, it is not a real fire test such as cone calorimeter. Therefore, it was decided to observe the real behavior of samples in front of direct flame using a set-up, which was adapted to the samples.

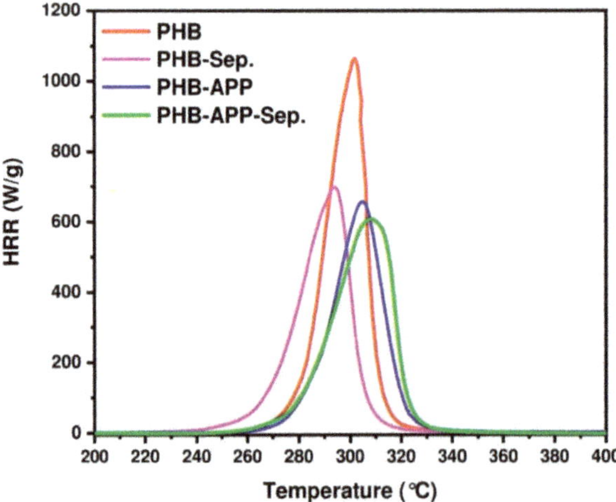

Figure 7. Heat Release Rate (HRR) curves obtained in Pyrolysis Combustion Flow Calorimeter PCFC tests.

Table 9. Summary results of Pyrolysis Combustion Flow Calorimeter (PCFC) tests.

Sample Code	THR (kJ/g)	T_{pHRR} (°C)	pHRR (W/g)	Reduction in pHRR (%)
PHB	22	302	1064	-
PHB-Sep.	17	291	656	38
PHB-APP	20	305	699	34
PHB-APP-Sep.	18	308	607	42

3.8. Vertical Burning Test

A vertical burning test was performed according to the method described earlier. Figure 8 shows the set-up used, while Figure 9 displays the digital photos taken at different times of test (All videos are available in Supplementary Materials). PHB was completely burned in 10 s and the time to which flame got dormant was 11.5 s. Incorporation of APP increased the time of flame out to 18 s and featured the autoextinguible behavior observed. The PHB-APP sample revealed high dripping behavior. The combination of APP and Sep. led to increase in time of flameout to 35.5 s; however, this sample completely burned with dripping behavior and without remaining residue. Surprisingly, the time of flame out significantly increased for the PHB-Sep. sample, reaching 59.7 s. A high amount of residue was remained at the end of the test and dripping behavior was again observed for this sample during the test. Except PHB-Sep., all samples showed dripping behavior during the flame test (see videos in Supplementary Materials, Video S1–S4). Although vertical test has not been recognized yet as a normalized standard fire test, our observations revealed that the behavior of different samples can be compared in terms of the time takes to inflammation and dripping. Moreover, by using this test the quantity and the integrity of final char of the PHB/Sep. sample was properly understood.

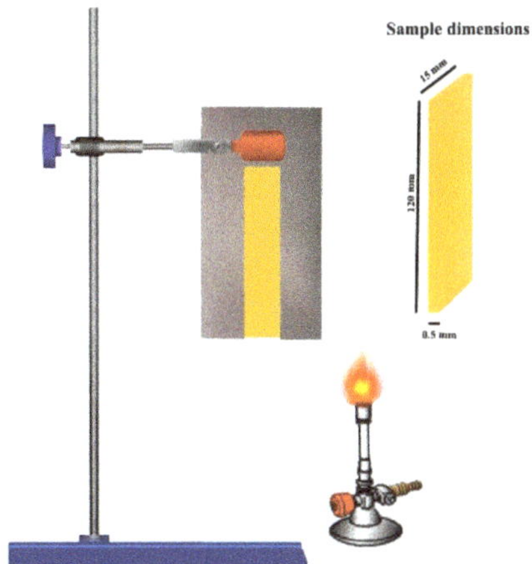

Figure 8. Illustration of flame test Set-up.

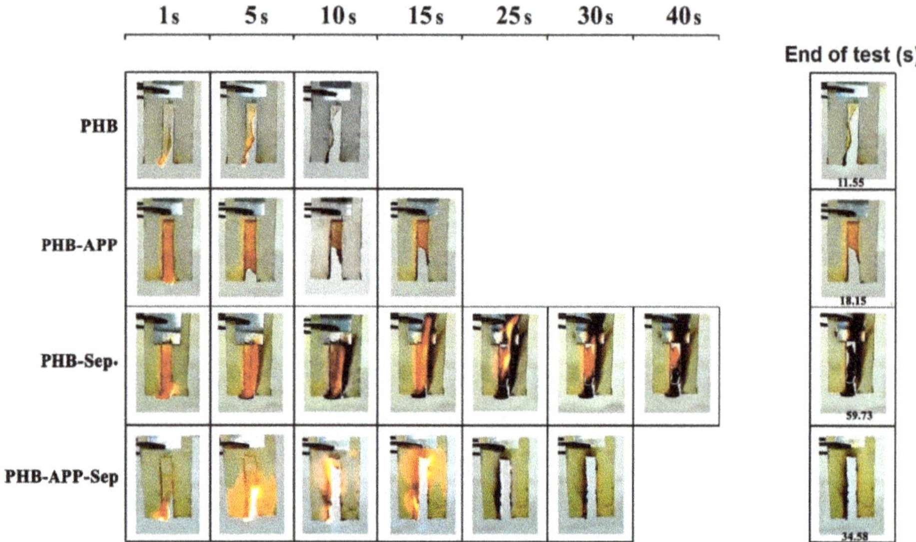

Figure 9. Evolution of flame propagation during vertical burning test for all samples.

4. Conclusions

The flammability and thermal stability behavior of PHB was investigated under the influence of the addition of organic ammonium polyphosphate (APP) as a conventional flame retardant, sepiolite as a inorganic biocompatible flame retardant, and lignin and starch. The preliminary assessment of flammability by PCFC confirmed that lignin and starch were the reason for high flammability of PHB. Then, investigations were continued by eliminating formulations in which lignin and starch were included. In the second phase, DSC results revealed that the presence of APP and sepiolite had no significant effect on glass transition and melting temperatures. TGA results showed that the presence of additives does not reduce the onset temperature of decomposition. In the case of combination of APP and sepiolite, the onset temperature was increased. The presence of sepiolite led to a significant increase in char residue, around 16 wt.% against 0 wt.% for pure PHB. PCFC results revealed the efficiency of sepiolite and its combination with APP to reduce peak of heat release rate, while vertical burning test clearly showed a huge difference behavior between PHB-Sep. and other samples. The peak of HRR of PHB was decreased from 1064 W/g to 656 W/g in presence of sepiolite (38% of reduction). The combination of sepiolite and APP was significantly affected the pHRR of PHB (42% of reduction).

Supplementary Materials: The following are available online at http://www.mdpi.com/1996-1944/12/14/2239/s1, Video S1: flame test-PHB; Video S2: flame test-PHB-APP; Video S3: flame test-PHB-APP-Sep.; Video S4: flame test-PHB-Sep.

Author Contributions: Conceptualization, H.V.; Formal analysis, H.V.; Investigation, H.V., L.M., G.M., V.A.; Methodology, H.V.; Supervision, H.V. and V.L.; Visualization, H.V., M.R.S., V.L., M.C., E.R. and C.V.; Writing–original draft, H.V., V.L., L.M.; Writing–review & editing, H.V., L.M., M.R.S. and V.L.

Funding: This research received no external funding.

Acknowledgments: The authors would like to thank "Plastinnov technological platform" for PCFC tests.

Conflicts of Interest: The authors declare no conflict of interest.

References

1. Nakajima, H.; Dijkstra, P.; Loos, K.J.P. The recent developments in biobased polymers toward general and engineering applications: Polymers that are upgraded from biodegradable polymers. Analogous to petroleum-derived polymers, and newly developed. *Polymers* **2017**, *9*, 523. [CrossRef]
2. Luckachan, G.E.; Pillai, C.K.S. Biodegradable polymers-a review on recent trends and emerging perspectives. *J. Polym. Environ.* **2011**, *19*, 637–676. [CrossRef]
3. Babu, R.P.; O'Connor, K.; Seeram, R. Current progress on bio-based polymers and their future trends. *Prog. Biomater.* **2013**, *2*, 2–8. [CrossRef] [PubMed]
4. Vahabi, H.; Rohani Rad, E.; Parpaite, T.; Langlois, V.; Saeb, M.R. Biodegradable polyester thin films and coatings in the line of fire: the time of polyhydroxyalkanoate (PHA). *Prog. Org. Coat.* **2019**, *133*, 85–89. [CrossRef]
5. Khosravi-Darani, K.; Bucci, D.Z. Application of poly (hydroxyalkanoate) in food packaging: Improvements by nanotechnology. *Chem. Biochem. Eng. Q.* **2015**, *29*, 275–285. [CrossRef]
6. Bugnicourt, E.; Cinelli, P.; Lazzeri, A.; Alvarez, V. Polyhydroxyalkanoate (PHA): Review of synthesis, characteristics, processing and potential applications in packaging. *Express Polym. Lett.* **2014**, *8*, 791–808. [CrossRef]
7. Cavaillé, L.; Albuquerque, M.; Grousseau, E.; Lepeuple, A.-S.; Uribelarrea, J.-L.; Guillermina, H.-R.; Paul, E. Understanding of polyhydroxybutyrate production under carbon and phosphorus-limited growth conditions in non-axenic continuous culture. *Bioresour. Technol.* **2016**, *201*, 65–73. [CrossRef] [PubMed]
8. Elvers, D.; Song, C.H.; Steinbüchel, A.; Leker, J. Technology Trends in Biodegradable Polymers: Evidence from Patent Analysis. *Polym. Rev.* **2016**, *56*, 584–606. [CrossRef]
9. Ariffin, H.; Nishida, H.; Shirai, Y.; Hassan, M.A. Determination of multiple thermal degradation mechanisms of poly(3-hydroxybutyrate). *Polym. Degrad. Stab.* **2008**, *93*, 1433–1439. [CrossRef]
10. Gallo, E.; Schartel, B.; Acierno, D.; Russo, P. Flame retardant biocomposites: Synergism between phosphinate and nanometric metal oxides. *Eur. Polym. J.* **2011**, *47*, 1390–1401. [CrossRef]
11. Li, Z.; Yang, J.; Loh, J.L. Polyhydroxyalkanoates: Opening doors for a sustainable future. *NPG Asia Mater.* **2016**, *8*. [CrossRef]
12. Marosi, G.; Szolnoki, B.; Bocz, K.; Toldy, A. Fire-retardant recyclable and biobased polymer composites. In *Novel Fire Retardant Polymers and Composite Materials*; Woodhead Publishing: Cambridge, UK, 2017; Volume 1, pp. 117–146.
13. Yeo, J.C.C.; Muiruri, J.K.; Thitsartarn, W.; Li, Z.; He, C. Recent advances in the development of biodegradable PHB-based toughening materials: Approaches, advantages and applications. *Adv. Mater. Sci. Eng.* **2018**, *92*, 1092–1116. [CrossRef] [PubMed]
14. Wu, W.; He, H.; Liu, T.; Wei, R.; Cao, X.; Sun, Q.; Li, R.K. Synergetic enhancement on flame retardancy by melamine phosphate modified lignin in rice husk ash filled P34HB biocomposites. *Compos. Sci. Technol.* **2018**, *168*, 246–254. [CrossRef]
15. Battegazzore, D.; Noori, A.; Frache, A. Hemp hurd and alfalfa as particle filler to improve the thermo-mechanical and fire retardant properties of poly(3-hydroxybutyrate-co-3-hydroxyhexanoate). *Polym. Compos.* **2019**. [CrossRef]
16. Zhang, R.; Huang, H.; Yang, W.; Xiao, X. Preparation and thermal properties of poly(3-hydroxybutyrate-co-4-hydroxybutyrate)/aluminum-containing layered double hydroxides nanocomposites. *High Perform. Polym.* **2013**, *25*, 104–112. [CrossRef]
17. Russo, P.; Carfagna, C.; Cimino, F.; Acierno, D.; Persico, P. Biodegradable composites reinforced with kenaf fibers: Thermal, mechanical, and morphological issues. *Adv. Polym. Tech.* **2013**, *32*, E313–E322. [CrossRef]
18. Gallo, E.; Schartel, B.; Acierno, D.; Cimino, F.; Russo, P. Tailoring the flame retardant and mechanical performances of natural fiber-reinforced biopolymer by multi-component laminate. *Compos. Part B: Eng.* **2013**, *44*, 112–119. [CrossRef]
19. Russo, P.; Vetrano, B.; Acierno, D.; Mauro, M. Thermal and structural characterization of biodegradable blends filled with halloysite nanotubes. *Polym. Compos.* **2013**, *34*, 1460–1470. [CrossRef]
20. Olmo, N.; Lizarbe, M.A.; Gavilanes, J.G. Biocompatibility and degradability of sepiolite-collagen complex. *Biomaterials* **1987**, *8*, 67–69. [CrossRef]

21. Masood, F.; Haider, H.; Yasin, T. Sepiolite/poly-3-hydroxyoctanoate nanocomposites: Effect of clay content on physical and biodegradation properties. *Appl. Clay Sci.* **2019**, *175*, 130–138. [CrossRef]

22. Ajmal, A.W.; Masood, F.; Yasin, T. Influence of sepiolite on thermal, mechanical and biodegradation properties of poly-3-hydroxybutyrate-co-3-hydroxyvalerate nanocomposites. *Appl. Clay Sci.* **2018**, *156*, 11–19. [CrossRef]

23. Vahabi, H.; Sonnier, R.; Otazaghine, B.; Le Saout, G.; Lopez-Cuesta, J.M. Nanocomposites of polypropylene/polyamide 6 blends based on three different nanoclays: Thermal stability and flame retardancy. *Polimery* **2013**, *58*, 350–360. [CrossRef]

24. Huang, N.H.; Chen, Z.J.; Wang, J.Q.; Wei, P. Synergistic effects of sepiolite on intumescent flame retardant polypropylene. *Express Polym. Lett.* **2010**, *4*, 743–752. [CrossRef]

25. Pappalardo, S.; Russo, P.; Acierno, D.; Rabe, S.; Schartel, B. The synergistic effect of organically modified sepiolite in intumescent flame retardant polypropylene. *Eur. Polym. J.* **2016**, *76*, 196–207. [CrossRef]

26. Zhang, M.; Thomas, N.L. Preparation and properties of polyhydroxybutyrate blended with different types of starch. *J. Appl. Polym. Sci.* **2010**, *116*, 688–694. [CrossRef]

27. Godbole, S.; Gote, S.; Latkar, M.; Chakrabarti, T. Preparation and characterization of biodegradable poly-3-hydroxybutyrate–starch blend films. *Bioresour. Technol.* **2003**, *86*, 33–37. [CrossRef]

28. Mousavioun, P.; Halley, P.J.; Doherty, W.O. Thermophysical properties and rheology of PHB/lignin blends. *Ind. Crops Prod.* **2013**, *50*, 270–275. [CrossRef]

29. Bertini, F.; Canetti, M.; Cacciamani, A.; Elegir, G.; Orlandi, M.; Zoia, L. Effect of ligno-derivatives on thermal properties and degradation behavior of poly (3-hydroxybutyrate)-based biocomposites. *Polym. Degrad. Stab.* **2012**, *97*, 1979–1987. [CrossRef]

30. Lyon, R.E.; Walters, R.N.; Stoliarov, S.I. Screening flame retardants for plastics using microscale combustion calorimetry. *Polym. Eng. Sci.* **2007**, *47*, 1501–1510. [CrossRef]

31. Cogen, J.M.; Lin, T.S.; Lyon, R.E. Correlations between pyrolysis combustion flow calorimetry and conventional flammability tests with halogen-free flame retardant polyolefin compounds. *Fire Mater.* **2009**, *33*, 33–50. [CrossRef]

32. Morgan, A.B. *Milligram Scale Flammability Testing of Flame Retardant Polyurethane Foams*; ACS Symposium Series; American Chemical Society: Washington, DC, USA, 2012; Volume 1118, pp. 445–458.

33. Sonnier, R.; Vahabi, H.; Ferry, L.; Lopez-Cuesta, J.M. *Pyrolysis-combustion Flow Calorimetry: A Powerful Tool to Evaluate the Flame Retardancy of Polymers*; ACS Symposium Series; American Chemical Society: Washington, DC, USA, 2012; Volume 1118, pp. 361–390.

34. Murariu, M.; Bonnaud, L.; Yoann, P.; Fontaine, G.; Bourbigot, S.; Dubois, P. New trends in polylactide (PLA)-based materials: "Green" PLA–calcium sulfate (nano) composites tailored with flame retardant properties. *Polym. Degrad. Stab.* **2010**, *95*, 374–381. [CrossRef]

35. Aoyagi, Y.; Yamashita, K.; Doi, Y. Thermal degradation of poly[(R)-3-hydroxybutyrate],poly[e-caprolactone], and poly[(S)-lactide]. *Polym. Degrad. Stab.* **2002**, *76*, 53–59. [CrossRef]

36. Li, S.D.; He, J.D.; Yu, P.; Cheung, M.K. Thermal degradation of poly(3-hydroxybutyrate) and poly(3-hydroxybutyrate-co-3-hydroxyvalerate) as studied by TG, TG–FTIR, and Py–GC/MS. *J. Appl. Polym. Sci.* **2003**, *89*, 1530–1536. [CrossRef]

37. Nguyen, S.; Yu, G.; Marchessault, R.H. Thermal Degradation of Poly(3-hydroxyalkanoates): Preparation of Well-Defined Oligomers. *Biomacromolecules* **2002**, *3*, 219–224. [CrossRef]

38. Gonzalez, A.; Irusta, L.; Fernandez-Berridi, M.J.; Iriarte, M.; Iruin, J.J. Application of pyrolysis/gas chromatography/Fourier transforminfrared spectroscopy and TGA techniques in the study ofthermal degradation of poly (3-hydroxybutyrate). *Polym. Degrad. Stab.* **2005**, *87*, 347–354. [CrossRef]

39. Boryniec, S.; Przygocki, W. Polymer Combustion Processes. 3. Flame Retardants for Polymeric Materials. *Prog. Rubber Plast. Recycl. Technol.* **2001**, *17*, 127–148. [CrossRef]

Article

Synergistic Effects of Aluminum Diethylphosphinate and Melamine on Improving the Flame Retardancy of Phenolic Resin

Ru Zhou [1,*], Wenjuan Li [1], Jingjing Mu [1], Yanming Ding [2] and Juncheng Jiang [1]

[1] Jiangsu Key Laboratory of Urban and Industrial Safety, College of Safety Science and Engineering, Nanjing Tech University, Nanjing 211816, China; 201761100038@njtech.edu.cn (W.L.); 201861201043@njtech.edu.cn (J.M.); jcjiang@njtech.edu.cn (J.J.)

[2] Faculty of Engineering, China University of Geosciences, Wuhan 430074, China; dingym@cug.edu.cn

* Correspondence: maxmuse.zhou@njtech.edu.cn

Received: 21 November 2019; Accepted: 30 December 2019; Published: 31 December 2019

Abstract: A series of novel flame retardants (aluminum diethylphosphinate and melamine) were used to improve the fire performance of phenolic resin. Fourier transform infrared spectroscopy (FTIR) was used to characterize the modification results. Thermo-gravimetric analysis (TGA) was used to study the thermal decomposition of phenolic resin system, and the flame retardancy of phenolic resin system was tested by vertical combustion test (UL-94) and limiting oxygen index (LOI). The combustion properties of modified phenolic resin were further tested with a cone calorimeter(CCT). Finally, the structure of carbon residue layer was measured by scanning electron microscopy (SEM). The results show that with the introduction of 10 wt % aluminum diethylphosphinate in phenolic resin, the LOI reaches 33.1%, residual carbon content increase to 55%. The heat release rate (HRR) decreased to 245.6 kW/m^2, and the total heat release (THR) decreased to 58.6 MJ/m^2. By adding 10 wt % aluminum diethylphosphinate and 3 wt % melamine, the flame retardancy of the modified resin can pass UL-94 V-0 flame retardant grade, LOI reaches 34.6%, residual carbon content increase to 59.5%. The HRR decreases to 196.2 kW/m^2 at 196 s, relatively pure phenolic resin decreased by 35.5%, and THR decreased to 51 MJ/m^2. Compared with pure phenolic resin, the heat release rate and total heat release of modified phenolic resin decreased significantly. This suggests that aluminum diethylphosphinate and melamine play a nitrogen-phosphorus synergistic effect in the phenolic resin, which improves the thermal stability and flame retardancy of the phenolic resin.

Keywords: phenolic resin; aluminum diethylphosphinate; melamine; flame retardancy

1. Introduction

With the development of world's social economy, high-rise buildings and skyscrapers have occupied a certain role. Regarding the country's vigorous promotion of building energy conservation, thermal insulation materials, especially organic building insulation materials, are widely used in building insulation construction due to their excellent thermal insulation properties, low cost, pressure and water resistance. The organic thermal insulation materials currently used for outside wall insulation include polystyrene foam board (EPS), extruded polystyrene foam board (XPS), polyurethane foam board (PU), and phenolic foam board [1,2].

One of the applications of phenolic resin as insulation material is to make phenolic foam insulation board. The phenolic resin can activate the interface by the surfactant and reduce the surface tension. The foaming agent can be decomposed into gas or physically vaporized to form a honeycomb structure, and finally the phenolic foam is formed under the action of curing and curing of the curing agent. Compared with other outside wall insulation materials, the phenolic foam has many advantages in thermal insulation and fire resistance, which is known as the "king of insulation" [3,4]. Since the 1990s, phenolic composite materials, including phenolic foam, have been significantly developed. Firstly, the military of the United Kingdom and the United States has paid attention to it. It has been used in the aerospace, defense and military industries, and later used in ships, stations, oil wells and other places with strict fire protection requirements and is gradually being promoted in construction, hospitals, stadiums and other fields [5]. In recent years, phenolic foam as the sound insulation and insulation materials have been widely used in the construction industry. Japan's Ministry of Construction has issued a decree on phenolic foams as standard building flame retardant materials, and France uses phenolic foam material as a material to seal and control fires. Many large apartments in cities such as Marseille and Lyon have phenolic foam panels installed on the walls. Ordinary phenolic resin is relatively brittle, and the phenolic hydroxyl group and methylene group in the molecular structure are easily oxidized, which limits the application of phenolic resin at high temperature. Therefore, the phenolic resin is further modified to enhance its fireproof and flame retardant properties so that a modified phenolic resin which is more suitable for the market can be obtained. Commonly used are phosphorus-based, nitrogen-based flame retardants, silicones, boron. Sang used graphene and cardanol to modify phenolic resin to get graphene and cardanol-modified phenolic resin (GCP). Filling the resin with homemade carbon fiber base paper to prepare graphene and cardanol, modified phenolic resin-based carbon fiber paper-based composite (GCPC) is created [6]. Tang and others used polyammonium phosphate (APP) as flame retardant to modify the phenolic resin composites reinforced by bamboo and polypropylene fiber composite felt. They studied the effects of adding flame retardant on the mechanical, flame retardant and thermal conductivity of the composites [7]. Li and others used borosilicate modified phenolic resin, boron and silicone to prepare a novel boron-and silicone-containing phenolic resin (BSiPF) solution, and characterized the structure, molecular weight, gel properties, curing properties and heat resistance of BSiPF. This makes introducing silicon and boron significantly improve the heat resistance and oxidation resistance of phenolic resin [8]. Wang used hydrobromic acid(HBr) and hydroiodic acid(HI) as reagents to demethylate an alkali lignin (AL) to increase its hydroxyl content and measured different thermal properties and performance of phenolic resins [9]. Yong and Yu used bio-oil as the raw material to synthesize bio-oil phenol formaldehyde resin desirable resin for fabricating phenolic-based material, which the thermal curing behavior and heat resistance of bio-oil phenol formaldehyde resins were investigated [10,11].

Melamine is a halogen-free nitrogen-based flame retardant, which has the characteristics of improving the physical and mechanical properties of products, low hygroscopicity, replacing some of the resin, reducing the amount of the adhesive. Compared to other types of flame retardants, melamine has apparent advantages. Melamine has been widely used in the modification of phenolic resins. Ge and others modified phenolic resin with melamine, and through experiments the mechanical properties of the modified phenolic foam were improved, while the free formaldehyde content decreased and the oxygen index increased. The research shows that the best melamine mass fraction is 4.5% [12]. Aluminum diethylphosphinate is a new flame retardant organic hypophosphite. It has excellent flame retardant properties, environmental and health-friendly properties, halogen-free, high flame-retardant efficiency, hydrophobic smoke suppression, and thermal stability. Now mainly used in glass fiber-reinforced polyamide 6 and 66 are composites and various polyesters (PBT and PET). Chen Lu introduced aluminum diethylphosphinate into polyethylene terephthalate (PET), and tested the flame retardancy of PET composites by vertical combustion and oxygen index. The thermal decomposition activation energy was calculated by thermo-gravimetric (TG) analyzer and the Flynn–Wall–Ozawa method, the thermal degradation behavior was further discussed [13]. Zhao employed four types of metal-mediated catalysts in the synthesis of phenolic resin resin to accelerate the curing rate and to lower the curing temperature of phenolic resin [14]. Chen and others prepared a series of polylactic acid/aluminum diethylphosphinate and polylactic acid/aluminum diethylphosphinate/organomontmorillonite composites, and analyzed them by thermo-gravimetric analysis, limiting oxygen index, and vertical combustion. The effects of aluminum diethylphosphinate and organomontmorillonite on the thermal stability and flame retardancy of polylactic acid composites were investigated by cone calorimeter. The mechanism of the synergistic flame-retardant character of aluminum diethylphosphinate and organomontmorillonite was also discussed [15].

Melamine flame retardants have been used in phenolic resins, but aluminum diethylphosphinate has not been used in phenolic resins. The two flame retardants are combined and added into phenolic resin as a new composite flame retardant. In this work, aluminum diethylphosphinate and melamine are added to the phenolic resin as excellent flame-retardant additives to improve the flame-retardant properties of phenolic resin. The nitrogen-phosphorus additive acts to make the modified phenolic resin more excellent. Through the test of its flame retardancy, thermal stability and combustion properties, the effects of the two additives on the phenolic resin were clarified, and the optimal synergistic ratio was sought, so phenolic resin could be more widely used in building insulation materials.

2. Materials and Methods

2.1. Experimental Materials

Phenolic resin, Shandong baiqian chemical Co., Ltd, Shandong, China. Aluminum diethylphosphinate, melamine, chemical pure, Sinopharm Chemical Reagent Co., Ltd, Nanjing, China. Anhydrous ethanol, Yonghua Chemical Technology (Jiangsu) Co., Ltd, Nanjing, China. All the raw materials were industrial products.

2.2. Preparation of Modified Phenolic Resin

Anhydrous ethanol was added to the phenolic resin, stirred to dissolve, and different ratios of aluminum diethylphosphinate and melamine were added and stirred with a constant temperature magnetic stirrer to dissolve it evenly. Then, this was subject to 30 min with an ultrasonic disperser. After standing for 20 min, the bubbles were removed, the mix was poured into a Teflon mold, and placed in a vacuum drying oven to step up the temperature. The specific addition amounts of aluminum diethylphosphinate and melamine are shown in Table 1.

Table 1. Samples of phenolic resin composite materials containing different additives.

Sample Codes	Phenolic Resin (wt %)	Aluminum Diethylphosphinate (wt %)	Melamine (wt %)
E-0	100	0	0
E-a1	95	5	0
E-a2	90	10	0
E-a3	85	15	0
E-a4	80	20	0
E-b1	98	0	2
E-b2	96	0	4
E-b3	94	0	6
E-b4	92	0	8
E-c1	87	10	3
E-c2	82	15	3
E-c3	86	10	4
E-c4	81	15	4
E-c5	85	10	5
E-c6	80	15	5

2.3. Testing and Characterization

Fourier transform infrared spectroscopy (FTIR): measurement was carried out with IS5 infrared spectrometer (Thermo Scientific, Waltham, MA, USA); thermo-gravimetric analysis (TGA) experiment: the spline was subjected to TG test using a SDTQ600 thermo-gravimetric analyzer (TA Instruments, Lukens, DE, USA). The temperature range was 50 °C–800 °C and it was tested under a nitrogen atmosphere with a gas flow rate of 20 mL/min; limiting oxygen index (LOI) test: the standard is ISO 4589-2 [16], and we used the HC-2 type limit oxygen index instrument (Jiangning Instrument Analysis Factory, Nanjing, China) was tested, the test sample size was $100 \times 10 \times 4$ mm^3; vertical combustion test(UL-94): a CFZ-2 horizontal vertical burner (Jiangning Instrument Analysis Factory, Nanjing, China) was used, and the test sample size was $125 \times 13 \times 3$ mm^3; cone calorimeter test (CCT): this employed the ISO 5660 standard [17], and a Govmark CC-2-2128 cone calorimeter (Deatak, McHenry, IL, USA)was used to test the combustion performance of phenolic resin and its composite materials in the atmosphere of 50 kW/m^2 heat flux. The size of the test sheet was $100 \times 100 \times 3$ mm^3. Scanning electron microscopy (SEM): the morphological structures of the char layer after the calorimeter test were observed by Zeiss EVO18 SEM, Oberkochen, Germany.

3. Results and Discussion

3.1. Fourier Transform Infrared Spectroscopy (FTIR) Test Results

Phenolic resin and modified phenolic resin were characterized by Fourier transform infrared spectroscopy, as shown in Figure 1. The Fourier spectra of the melamine-modified phenolic resin are shown in Figure 1, and the absorption peaks at 3469 cm^{-1} and 3419 cm^{-1} are the performance of the anti-symmetrical vibration of NH$_2$. The 1022 cm^{-1} is formed by NH twisting vibration, and the characteristic absorption peak at 814 cm^{-1} is the characteristic absorption of deformation vibration of a triazine ring [18]. The absorption peaks of these bands are also found in Figure 1 of the phenolic resin system in which aluminum diethylphosphinate and melamine are simultaneously added. In Figure 1 of the aluminum diethylphosphinate modified phenolic resin system, the stretching vibration of P=C double bond is at 1153 cm^{-1}, and the peak at 1095 cm^{-1} is formed by PO· symmetrical stretching vibration [19]. The characteristic peak at 2955 cm^{-1} is consistent with the stretching vibration frequency of –CH$_2$– [20], and characteristic peak coincides with the characteristic peak of aluminum diethylphosphinate. The phenolic resin system with aluminum diethylphosphinate and melamine added at the same time has these absorption peak. In addition, the other characteristic peaks are consistent with the pure phenolic resin. The results were confirmed that the aluminum diethylphosphinate and melamine were synthesized successfully.

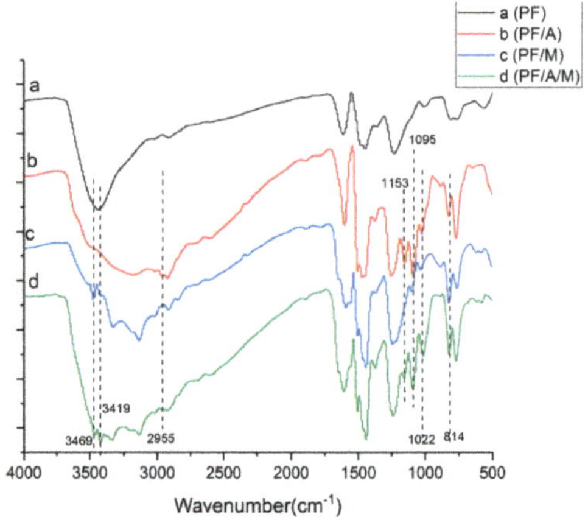

Figure 1. Fourier transform infrared (FTIR) spectra of phenolic resin and phenolic resin composite system.

3.2. Thermo-Gravimetric Analysis (TGA) Results

3.2.1. TGA of Modified Phenolic Resin

The thermal degradation performance of the modified phenolic resin was analyzed by the thermo-gravimetric test, and the melamine phenolic resin composite and the thermal degradation process under the nitrogen atmosphere. Figures 2 and 3 are TGA curves of aluminum diethylphosphinate modified phenolic resin and melamine modified phenolic resin, respectively. Figure 2 is the aluminum diethylphosphinate/phenolic resin composite in a temperature range of 50 °C to 800 °C under a nitrogen atmosphere [21]. The results are shown in Table 2. $T_{5\%}$, $T_{10\%}$ and T_{dmax} in Table 2 stand for the temperature at 5 wt %, 10 wt % and maximum weight loss rate, respectively. The thermal degradation of pure phenolic resin under the nitrogen atmosphere is a one-step degradation process. When the temperature reaches 178 °C, it begins to decay. When the temperature reaches 550 °C, the thermal decomposition rate is the fastest, and carbon residue at 800 °C is 51 wt %. Compared with pure phenolic resin, the TGA curves of aluminum diethylphosphinate/phenolic resin composites are different. When the addition amount reaches 15%, two degradation processes occur at 465 °C and 560 °C, respectively. The T_{dmax} of E-0, E-a1, E-a2, E-a3 and E-a4 are 550, 562, 557, 561 and 467 °C respectively. The maximum thermal weight loss rates for E-0, E-a1, E-a2, E-a3 and E-a4 are 0.174, 0.191, 0.167, 0.194 and 0.203%/°C. The thermal stability of aluminum diethylphosphinate/phenolic resin composites is significantly different compared with pure phenolic resin. The figure shows that when the added amount of aluminum diethylphosphinate is about 10%, residual carbon of the composite at 800 °C is the largest, which is 55 wt %. When the amount of aluminum diethylphosphinate is increased to 20%, the residual carbon rate is reduced, the maximum mass loss rate becomes larger. This is because aluminum diethylphosphinate acts as a flame retardant through the gas phase flame retardant mechanism, and when the temperature rises, aluminum diethylphosphinate degrades the phosphorus-containing compound. During combustion, it is decomposed into diethylphosphonic acid, which volatilizes to the gas phase and inhibits combustion by capturing free radicals produced by combustion [22]. However, when added in a large amount, aluminum diethylphosphinate will reduce the thermal stability of phenolic resin. Excessive aluminum diethylphosphinate addition may affect the molecular properties of the phenolic resin, thus reducing the flame retardancy and thermal stability of the phenolic resin. Therefore, the aluminum diethylphosphinate should be in the range of 10%~15%, which cannot only ensure thermal stability, but also perform its flame-retardant carbon properties.

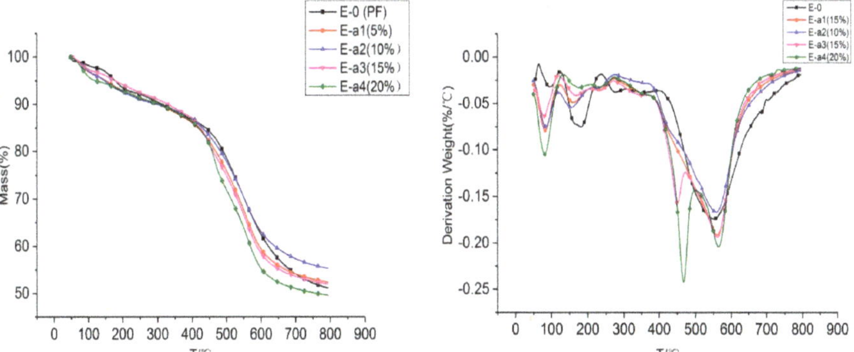

Figure 2. Thermo-gravimetric (TG) and derivative thermo-gravimetric (DTG) images of aluminum diethylphosphinate modified phenolic resin under nitrogen atmosphere.

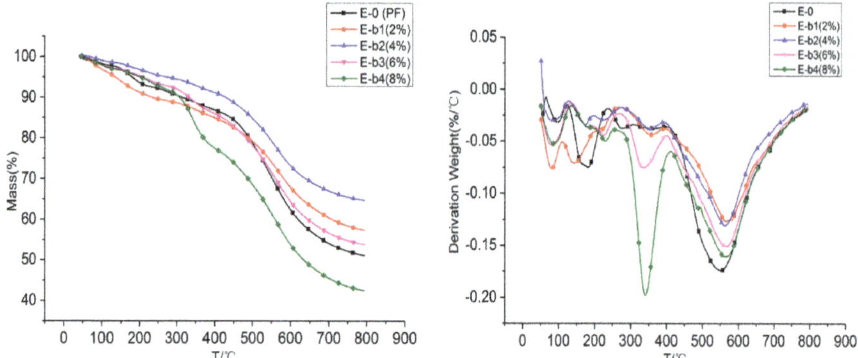

Figure 3. TG and DTG images of melamine modified phenolic resin under nitrogen atmosphere.

Table 2. TG data of phenolic resin and its composites.

Material	$T_{5\%}$ (°C)	$T_{10\%}$ (°C)	T_{dmax} (°C)	Char Residues at 800 °C (wt%)
E-0	178	313	550	51
E-a1	151	300	562	52
E-a2	147	299	557	55
E-a3	177	329	561	52
E-a4	121	305	467	49

Figure 3 is the TGA curve of phenolic resin and melamine phenolic resin composite material in the temperature range of 50–800 under the nitrogen atmosphere [23]. The results are shown in Table 3. $T_{5\%}$, $T_{10\%}$ and T_{dmax} in Table 3 stand for the temperature at 5 wt %, 10 wt % and maximum weight loss rate, respectively. The TGA curve trend of E-b1, E-b2, E-b3 and E-b4 samples is similar to that of pure phenolic resin, which is mainly composed of two thermogravimetry stages. From Figure 3, we can see that the thermal degradation of pure phenolic resin under the nitrogen atmosphere is a one-step degradation process. When the temperature reaches 178 °C, it begins to decay. When the temperature reaches 550 °C, the thermal weight loss rate of pure phenolic resin reached the maximum at 550 °C and is 0.174 %/°C. By adding melamine into phenolic resin, the curves are different. The maximum thermal weight loss rates corresponding to E-b1, E-b2 and E-b3 are 0.126, 0.13 and 0.151%/°C, which are all lower than 0.174 %/°C of the E-0 groups. When the addition of 8% melamine to phenolic resin, the thermal weight loss rate of E-b4 is different to the other group, and its maximum thermal weight loss rate is 0.197%/°C at 340 °C. A peak appeared between 270 °C to 410 °C, and the peak value increased with the increase of the amount of melamine added. The decomposition stage at this stage is due to the decomposition of melamine. The second thermal weight loss phase occurs between 410 °C –650 °C, mainly due to the fracture of the phenolic resin main chain caused. At this stage, more obvious peaks can be seen from the DTG curve. This is because with the addition of melamine, the thermal weight loss rate of the composite gradually decreases. The melamine phenolic resin composite has a $T_{10\%}$ ratio of 231 °C, 426 °C, 331 °C and 306 °C, and the residual carbon content at 800 °C is 57%, 64%, 53%, 42%. Thus it can be seen that with the addition of 4% melamine to phenolic resin, the residual carbon content reached the maximum, and the maximum mass loss rate was also small. This shows that the thermal stability of the melamine phenolic resin composite was better at 4% addition, and there was residual in the synthesis of phenolic resin. There is residual formaldehyde in the synthesis of phenolic resin, and formaldehyde is flammable. With the addition of melamine, it reacts with formaldehyde to form melamine formaldehyde, which consumes a part of residual formaldehyde. Moreover, the generated melamine formaldehyde contains nitrogen rings, which have a good flame retardant effect [24], thus increasing the thermal stability of phenolic resin.

Table 3. TG data of phenolic resin and its composites.

Material	$T_{5\%}$ (°C)	$T_{10\%}$ (°C)	T_{dmax} (°C)	Char Residues at 800 °C (wt %)
E-0	178	313	550	51
E-b1	133	231	559	57
E-b2	268	426	560	64
E-b3	206	331	566	53
E-b4	198	306	340	42

3.2.2. TGA of Phenolic Resin Modified by Synergistic Modification of Aluminum Diethylphosphinate and Melamine

The thermo-gravimetric analysis curve was used to study the thermal stability of the aluminum diethylphosphinate and melamine synergistic modified phenolic resin composites and the thermal degradation process. Figure 4 is the TGA curve of a phenolic resin and its composite material in a temperature range of 50 °C to 800 °C under a nitrogen atmosphere. Some results are shown in Table 4. $T_{5\%}$, $T_{10\%}$ and T_{dmax} in Table 4 stand for the temperature at 5 wt %, 10 wt % and maximum weight loss rate, respectively. According to Figure 3, the TG curves of the six groups of E-c1, E-c2, E-c3, E-c4, E-c5 and E-b6 have similar trends, mainly composed of two thermo-gravimetric stages. The T_{dmax} values of the phenolic resin composite system were 570 °C, 560 °C, 559 °C, 559 °C, 559 °C, and 561 °C, respectively. Under the same range of temperatures, the residual carbon content at 800 °C was 59.5%, 50.7%, 57.7%, and 52.2%, 54.3%, 49.5%. When adding 3 wt % melamine and 10 wt % aluminum diethylphosphinate, the residual carbon amount is up to 59.5%, adding 10 wt % aluminum diethylphosphinate alone, the residual carbon amount is 55%. The carbon residue rose when the melamine was added to phenolic resin, which means that melamine can greatly improve the thermodynamic stability of phenolic resin. When the amount of aluminum diethylphosphinate and melamine added are 15% and 5% by weight, the residual carbon is 49.5%. Compared with the pure phenolic resin, the stability has not been improved because the flame retardant is excessive, which causes the thermodynamic properties of the phenolic resin to be negatively affected.

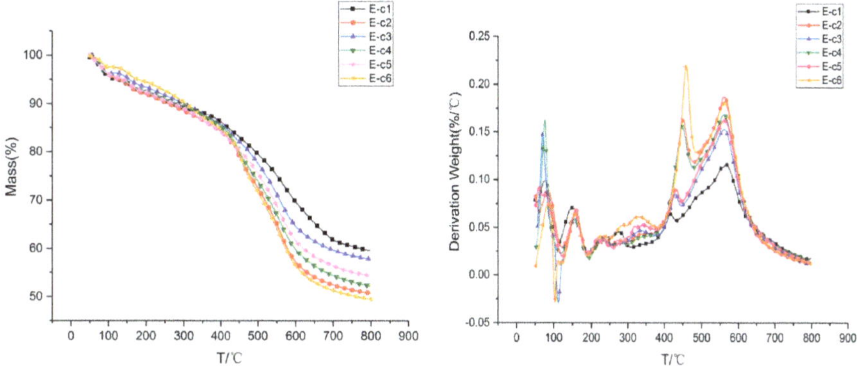

Figure 4. TG and DTG images of aluminum diethylphosphinate and melamine modified phenolic resin under nitrogen atmosphere.

Table 4. TG data of phenolic resin composites.

Material	$T_{5\%}$ (°C)	$T_{10\%}$ (°C)	T_{dmax} (°C)	Char Residues at 800 °C (wt %)
E-c1	121	265	570	59.5
E-c2	125	256	560	50.7
E-c3	155	299	559	57.7
E-c4	137	281	559	52.2
E-c5	138	271	559	54.3
E-c6	176	303	561	49.5

3.3. Modified Phenolic Resin Flame-Retardant Properties

The flame-retardant properties of different resin systems were tested by LOI and the UL-94 vertical burning test [25]. Tables 5 and 6 presents the LOI values and UL-94 test results of the samples.

Table 5. Limiting oxygen index (LOI) and vertical burning test (UL-94) results of phenolic resin and its composites.

Sample Codes	Flame Retardants Mass Fraction(%)	LOI	UL-94
Aluminum Diethylphosphinate			
E-0	0	30.0	V-2
E-a1	5	31.0	V-1
E-a2	10	33.1	V-0
E-a3	15	34.6	V-0
E-a4	20	32.3	V-0
Melamine			
E-b1	2	30.8	V-1
E-b2	4	32.1	V-0
E-b3	6	31.6	V-0
E-b4	8	30.4	V-1

Table 6. LOI and UL-94 results of phenolic resin composites.

Sample Codes	Aluminum Diethylphosphinate(wt %)	Melamine (wt %)	LOI	UL-94
E-c1	10	3	34.6	V-0
E-c2	15	3	34.8	V-0
E-c3	10	4	35.8	V-0
E-c4	15	4	34.4	V-0
E-c5	10	5	34.1	V-0
E-c6	15	5	34.0	V-0

According to Table 5, we can seen that the LOI value of pure phenolic resin is 30.0%, which is inherently flame-retardant, and almost no dripping is formed in the vertical burning test. Compared with that of pure phenolic resin, the LOI was 33.1% when the amount addition reached 10% by weight, and passed the UL-94 V-0 level in the vertical burning test. Continuing to increase the amount of flame retardant, when the addition amount reaches 15 wt%, the LOI of phenolic resin composites is 34.6%, and in the vertical burning test, can pass the UL-94 V-0 level. However, when the aluminum diethylphosphinate was 20 wt %, the flame-retardant capacity of phenolic resin decreased with the increase of the aluminum diethylphosphinate. It can be seen that when the addition amount of aluminum diethylphosphinate is 15%, the LOI and UL-94 test results of the phenolic resin composite system are the best, which can improve the flame-retardant performance of the phenolic resin. This highlights that most of the decomposition products of aluminum diethylphosphinate are volatilized into the gas phase during combustion, and may produce free radicals such as PO· in the gas phase, and capture free radicals produced by combustion to inhibit combustion [26,27]. When melamine was added to the phenolic resin, the LOI increased first and then decreased with the addition amount.

When the addition reached 4 wt %, the LOI reached a maximum of 32.1%. In the vertical burning test, it can pass the UL-94 V-0 level. Then, with the increase of the amount of addition, the LOI began to decrease, showing that adding a suitable amount of melamine can improve the flame-retardant properties of the phenolic resin.

The flame-retardant properties of aluminum diethylphosphinate and melamine were simultaneously added to the phenolic resin, as shown in Table 6. When the added amount of aluminum diethylphosphinate is 10 wt % and melamine is 4 wt %, the LOI of the phenolic resin composite system reaches a maximum value of 35.8%, and can pass the UL-94 V-0 grade. Compared with adding melamine alone, there is a significant improvement. With the addition of aluminum diethylphosphinate into phenolic resin, the phosphorus-containing group is thermally oxidized and decomposed during combustion to form phosphoric acid-promoting charcoal, suggesting that aluminum diethylphosphinate and melamine can play a synergistic role in the phenolic resin composite system. The synergistic effect of the nitrogen-phosphorus ion improves the flame retardancy. When the added amount of the two flame retardants is increased, the LOI value is decreased, indicating the addition amount has exceeded the appropriate range. If too much flame retardant is added, the flame retardancy of the phenolic resin system is lowered.

3.4. Cone Calorimeter Test (CCT) of Modified Phenolic Resin

The cone calorimeter test (CCT) can test the combustion properties of composites under real fire conditions, and it is one of the most important tests to study the flame retardant properties of composites [28]. The matching results of CCT were summarized in Table 7, including peak heat release rate (p_{HRR}), time corresponding to pHRR (T_{pHRR}), time to sustained ignition (TTI), total heat release (THR) and amount of carbon residue [29].

Table 7. Cone calorimeter test (CCT) data of phenolic resin and its composites.

Sample Codes	TTI (s)	P_{HRR} (kW/m^2)	T_{PHRR} (s)	THR (MJ/m^2)	Residues (wt %)
E-0	26	304.4	86	78.6	32.2
E-b2	29	286.6	90	62.7	36.7
E-a2	34	245.6	104	58.6	36.3
E-c6	38	213.9	158	55.3	37.9
E-c1	42	196.2	196	51	38.5

The HRR, THR, and mass loss plots of phenolic resin and its composites during combustion are shown in Figure 5. As can be seen from Figure 5 and Table 7, the HRR of pure phenolic resin increases sharply after ignition and reaches a peak heat release rate of 304.4 kW/m^2 at 86 s, and THR was as high as 78.6 MJ/m^2, TTI was 26 s, and the carbon residue was 32.8%. This shows that pure phenolic resin has a high risk of burning. When 10 wt % aluminum diethylphosphinate was added to the phenolic resin, the HRR of the E-a2 group decreased to 245.6 kW/m^2, the time was extended to 104 s, and the THR decreased to 58.6 MJ/m^2. Compared with pure phenolic resin, HRR and THR decreased by 19.3% and 25.4%, respectively. TTI increased to 34 s, and increased by 8 s compared with pure phenolic resin. After adding 10 wt % aluminum diethylphosphinate and 3 wt % melamine to the phenolic resin, the HRR of the E-c1 group decreased to 196.2 kW/m^2 and the THR decreased to 51 MJ/m^2. Compared with the pure phenolic resin, the HRR decreased by 35.5% and 35.1%, respectively. The TTI increased to 42, and the pure phenolic resin increased by 16 s. From the mass loss image of Figure 5c, it can be seen clearly that the mass loss curve of E-c1 is much more backward than pure phenolic resin. Compared with the pure phenolic resin residual carbon content of 32.2 wt %, the residual carbon content of E-c1 group is increased by 19.6%, which is 38.5 wt %. The simultaneous addition of aluminum diethylphosphinate and melamine is more effective than the phenolic resin system with only aluminum diethylphosphinate added. The simultaneous action of the two flame retardants can effectively improve the stability of the phenolic resin during combustion, reduce the heat release rate, heat released and the amount of residual carbon increase. Aluminum diethylphosphinate mainly produces flame retardant effect through the gas phase, while melamine can increase the amount of residual carbon and reduce the heat release rate.

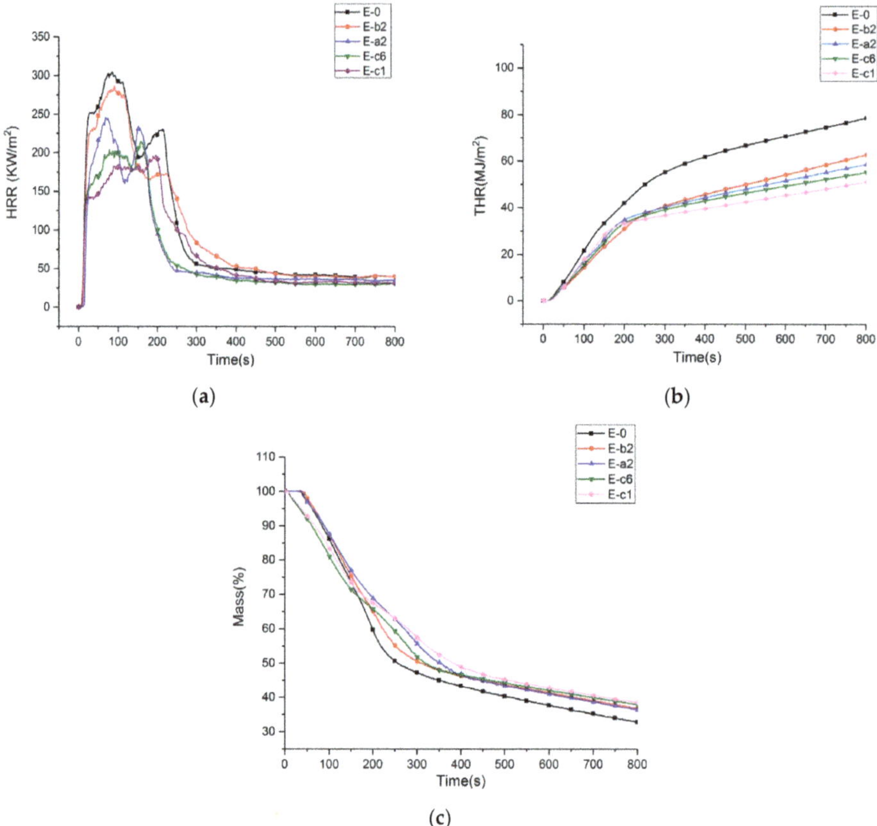

Figure 5. Heat release rate (**a**), total heat release (**b**) and weight loss curve (**c**) of phenolic resin and its composites.

By adding 4 wt % melamine to the phenolic resin, the HRR of the E-b2 group decreased to 286.6 kW/m^2, and the time was extended to 90 s. Compared with pure phenolic resin, HRR and THR decreased by 5.8% and 20.2%, respectively. The TTI increased to 29 s. Compared with the phenolic resin system with aluminum diethylphosphinate and melamine added at the same time, the thermal stability was obviously reduced when burning. When 15 wt % aluminum diethylphosphinate and 5 wt % melamine were added to the phenolic resin, the HRR reached a maximum of 213.9 kW/m^2 at 158 s, and THR values could reach 55.3 MJ/m^2. Compared with pure phenolic resin, the pHRR and THR of E-c6 decreased by 29.7% and 29.6%, respectively. It can be seen that compared with the E-c1 group phenolic composite material, the combustion performance of the E-c6 composite material was not good, which means that the excessive flame-retardant combustion performance is not effective.

According to research on the causes of death of those killed in fires, about two-thirds of deaths are caused by the direct poisoning of CO generated in the fire. Although CO_2 is non-toxic, when it reaches a certain concentration, it will also cause harm to the human body. Therefore, it is particularly important to reduce the content of CO and CO_2. Figure 6 shows CO and CO_2 curves of all composites; the results are summarized in Table 8, including peak concentration of CO (p_{CO}) and CO_2 (p_{CO2}), pink time of CO(Tp_{CO}) and CO_2 (Tp_{CO2}). The CO of pure phenolic resin increased sharply after ignition and reached the peak of 0.215% at 206 s, and CO_2 concentration was as high as 0.356% at 82 s. When 10 wt % aluminum diethylphosphinate was added to the phenolic resin, the maximum concentration of CO and CO_2 of the E-a2 group decreased to 0.201% at 152 s and 0.341%, respectively. Compared with pure phenolic resin, they decreased by 6.5% and 9%, respectively. By adding 10 wt % aluminum diethylphosphinate and 3 wt % melamine to the phenolic resin, the peak concentration of CO and CO_2 of the E-c1 group decreased to 0.196% and 0.222%, respectively. Compared with the pure phenolic resin, the content decreased by 8.8% and 37.6%, respectively. With the addition of 15 wt % aluminum diethylphosphinate and 5 wt % melamine to the phenolic resin, the concentration of CO and CO_2 of the E-c1 group decreased to 0.199% and 0.264%, respectively. Compared with the pure phenolic resin, they decreased by 7.4% and 0.134%, respectively. It is obvious that the CO and CO_2 concentrations were the minimum by adding 10 wt % aluminum diethylphosphinate to the phenolic resin, the hazard of the phenolic resin can be reduced.

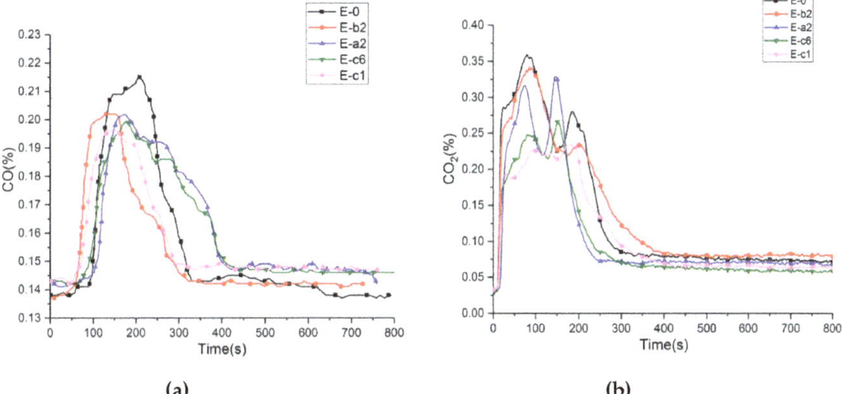

Figure 6. CO (a) and CO_2 (b) curves of phenolic resin and its composites.

Table 8. Gas data of phenolic resin and its composites.

Sample Codes	P_{CO} (%)	Tp_{CO} (s)	P_{CO2} (%)	Tp_{CO2} (s)
E-0	0.215	206	0.356	82
E-b2	0.202	152	0.341	90
E-a2	0.201	175	0.324	150
E-c6	0.199	183	0.264	154
E-c1	0.196	160	0.222	138

3.5. Analysis of Char Residual

To further understand the flame-retardant effect of the modified phenolic resin, the residual char after the cone calorimeter test was analyzed. Figures 7 and 8 are digital photographs of phenolic resin and composite materials before and after combustion under the same thermal radiation conditions in a cone calorimeter test.

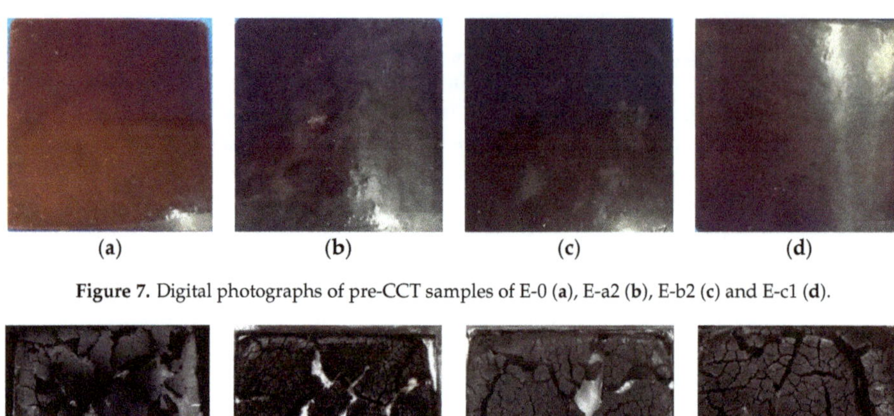

(a)   (b)   (c)   (d)

Figure 7. Digital photographs of pre-CCT samples of E-0 (**a**), E-a2 (**b**), E-b2 (**c**) and E-c1 (**d**).

(a)   (b)   (c)   (d)

Figure 8. Digital photographs of CCT samples after E-0 (**a**), E-a2 (**b**), E-b2 (**c**) and E-c1 (**d**).

As can be seen from the figures, pure phenolic resin is completely burned, and only a few carbon residues were left, and the residual carbon is obviously loose, light and thin. However, the carbon layer of E-a2 and E-b2 composites has a significant degree of expansion and an increase in the amount of carbon residue, as shown in Table 9. The thicknesses of the residual carbon layers in the E-0, E-a2, E-b2 and E-c1 groups were 2.5, 5.5, 5 and 5.8 mm, respectively. Compared with pure phenolic resin, the thickness of the residual carbon layer increased significantly. When aluminum diethylphosphinate and melamine are added together, the carbon residue of E-c1 is significantly increased compared to pure phenolic resin. The results show that the nitrogen-phosphorus ion obviously increases the amount of carbon residue and the strength of carbon layer after phenolic resin combustion.

Table 9. Thickness of residual carbon layer.

Sample Codes	Thickness of Residual Carbon Layer (mm)
E-0	2.5
E-a2	5.5
E-b2	5
E-c1	5.8

In order to explore the microstructure of the carbon layer of the phenolic composite, the carbon layer after the combustion of the pure phenolic resin and phenolic resin composites was analyzed by SEM. It can be observed from Figure 9a that the pure phenolic resin had a relatively porous hole and a large crack on the surface of the carbon layer after the CCT test, which is attributed to the rapid volatilization of the flammable volatile gas produced by the combustion on the pure phenolic resin surface. With the addition of aluminum diethylphosphinate, as shown in Figure 9b, the carbon layer of the phenolic resin composite system still had large pores, and the carbon layer was rather loose, which did not block oxygen and heat. This is because aluminum diethylphosphinate generates volatile diethyl hypophosphorous acid during degradation, which acts as a trapping free radical in the gas phase and blocks the combustion process. With the volatilization of diethyl hypophosphorous acid, the phosphorus content inside the phenolic resin substrate decreases rapidly, resulting in a decrease in the char forming ability of the system. Therefore, the effect of the condensed phase flame-retardant mechanism is not obvious [30], which can be explained through TGA. Therefore, the final residual amount of phenolic composite system with aluminum diethylphosphinate at 800 °C in TGA analysis has a small increase compared with that of pure phenolic resin material. When the phenolic system was introduced into melamine, as shown in Figure 9c, the carbon layer of the phenolic composite material did not have pores and cracks such as pure phenolic resin, and became smoother and finer. This is because melamine reacts with formaldehyde after it is added. The formation of melamine formaldehyde and melamine formaldehyde disturbs the intermolecular rigid structure of phenolic resin, and forms an irregular cross-linked interpenetrating network structure with phenolic resin, which hinders the decomposition of phenolic resin and needs higher energy in decomposition [31]. It can be explained that in the TGA analysis the final carbon residue of the phenolic composite system with melamine introduced at 800 °C was higher than that of the pure phenolic resin material. The carbon layer of the phenolic system with aluminum diethylphosphinate and melamine added at the same time, as shown in Figure 9d, was more compact, continuous and dense, and there were almost no large cracks and holes. This shows that the combined action of the two can play a good role in heat insulation and oxygen insulation, and can slow down the mass and heat transfer and the release of combustible volatiles, so as to prevent the further combustion of phenolic resin composite materials.

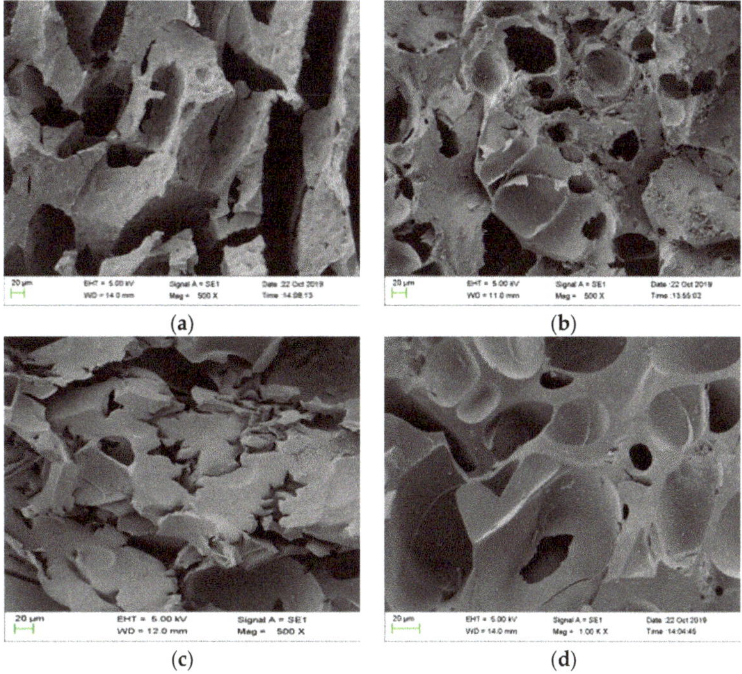

Figure 9. Electron micrograph of carbon layer after CCT test of phenolic resin and phenolic resin composites: E-0 (**a**), E-a2 (**b**), E-b2 (**c**) and E-c1 (**d**).

The FTIR spectra of the char residues after the cone calorimeter test are shown in Figure 10. By comparing with the spectra of the Figure 1, it can be seen that the peaks at 3469 cm^{-1} and 3419 cm^{-1} (NH$_2$ groups), 2955 cm^{-1} (–CH$_2$ groups), 1153 cm^{-1} (P=C groups) disappear, and some new peaks are formed during combustion, as shown in Figure 10, indicating that the cross-linking reaction occurred between the phenolic resin and two flame retardants. The strong characteristic peaks in the range of 3500–3100 cm^{-1} are attributed to the stretching vibration of –OH and –NH– groups. The peaks at 1632, 1400 and 997 cm^{-1} are due to the C=C absorption in carbonization reaction, P-N groups and P–O–C groups, respectively. After the addition of aluminum diethylphosphinate, a new peak is observed at 1247 cm^{-1} (P=O groups) and 781 cm^{-1} (Al–O groups) in Figure 10. Additionally, the char rich in P–O–P groups, P-O-C groups, P-N groups and C=C groups exhibits a more intumescent and compact structure, which gives an illustration of the lower heat release and gas emissions during the combustion process.

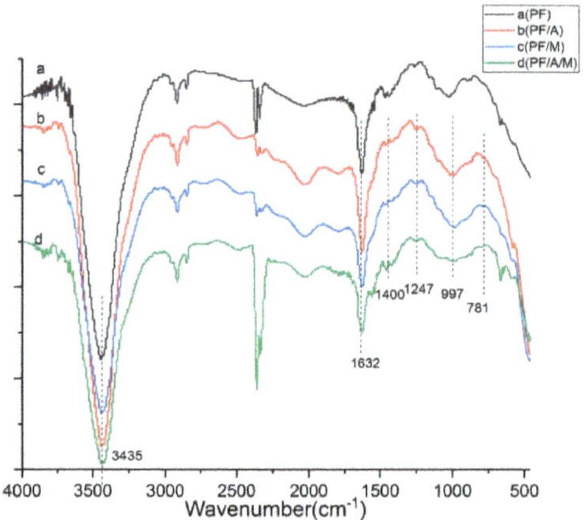

Figure 10. FTIR spectra of phenolic resin and phenolic resin composite system after the cone calorimeter test.

4. Conclusions

In this work, various concentrations of aluminum diethylphosphinate and melamine were successfully added to the phenolic resin to synthesize the modified phenolic resin. Through thermo-gravimetric analysis and the test of flame retardancy, the suitable addition amount of the two flame retardants was obtained. The results show that when 15 wt % of aluminum diethylphosphinate is added to the phenolic resin, the flame retardancy of the modified resin can pass the UL-94 V-0 flame-retardant grade, the LOI reaches 34.6%, and the residual carbon content is 52%. With the addition of 4 wt % melamine, the flame retardancy of the modified resin can pass the UL-94 V-0 flame retardant grade, the LOI reaches 32.1%, and the residual carbon amount is 64%. It can be seen that aluminum diethylphosphinate has a significant effect in improving the flame retardancy of the phenolic resin, and can increase LOI of the phenolic resin to 34.6%, while melamine can improve the thermal stability of phenolic resin, improve the carbon residue and reduce the rate of thermal decomposition. The test results of the cone calorimeter also show that when the aluminum diethylphosphinate and melamine are added together in the phenolic resin, the combustion performance is better than the addition of one of the flame retardants alone. With the addition of 10 wt% aluminum diethylphosphinate and 3 wt % melamine into the phenolic resin, the flame retardancy of the modified resin can pass the UL-94 V-0 flame-retardant grade, and the LOI reaches 34.6%. The residual carbon content increases to 59.5%, and the HRR decreases to 196.2 kW/m^2 at 196 s, which is relatively pure phenolic resin decreased by 35.5%, THR decreased to 51 MJ/m^2 and TTI increased to 42 s. The peak content of CO and CO$_2$ decreased to 0.196% and 0.222%, compared with the pure phenolic resin, and the content decreased by 8.8% and 37.6%, respectively. In the SEM experiment, the electron micrograph of residual carbon also verified that the carbon layer of the phenolic system with aluminum diethylphosphinate and melamine added at the same time is more compact, continuous and dense, while there are almost no large cracks and holes, indicating that the two flame retardants can work well together. It acts as a heat insulator and oxygen barrier to slow the release of mass and heat transfer and flammable volatiles, thereby preventing further combustion of the phenolic resin composite. All the above suggest that aluminum diethylphosphinate and melamine play a nitrogen-phosphorus synergistic effect in the phenolic resin, which improves the thermal stability and flame retardancy of the phenolic resin. The thermal stability and flame retardancy of phenolic resin obtained can be used in further tests on pyrolysis and combustion in building materials.

Author Contributions: Conceptualization, R.Z.; data curation, W.L. and J.M.; project administration, Y.D. and J.J. All authors have read and agreed to the published version of the manuscript.

Funding: This research was funded by National Key Research and Development Program of China (No.2018YFC0809300), National Natural Science Foundation of China (51874183) and Natural Science Fund of the Jiangsu Higher Education Institutions of China (No. 18KJA620004).

Conflicts of Interest: The authors declare no conflict of interest.

References

1. Yi, A.; Liu, J. Study on the thermal decomposition of organic heat insulation materials. *New Chem. Mater.* **2011**, *39*, 94–96. (In Chinese)
2. Wang, W.; Ye, Y.; Yu, Q. Research progress in thermal degradation behaviors of organic thermal insulation materials. *Mater. Rev.* **2014**, *28*, 100–103. (In Chinese)
3. Yao, J.; Hou, Z.; Yao, Y.S. Application of phenolic foam plate in the exterior wall thermal insulation. *Appl. Mech. Mater.* **2012**, *174–177*, 1363–1366. [CrossRef]
4. Sun, Z.; Wang, L.; Li, D. Study of preparation and property of phenolic foam. *China Plast. Ind.* **2007**, *35*, 46–49.
5. Yin, Y.; Liu, H. Fireproof and thermal isolation being the required performance for energy-saving insulation buildings—discussion on the development trend of phenolic foam. *Constraction Sci. Technol.* **2019**, *362*, 70–74.
6. Sang, M.; Wang, F.; Wang, S.-H. Preparation and effect of graphene modified cardanol-aldehyde resin on carbon fiber paper. *China Plast. Ind.* **2018**, *46*, 67–71.
7. Tang, Q.; Lu, F.; Guo, W. Effects of ammonium polyphosphat on properties of bamboo/polypropylene mats reinforced phenolic resin composites. *China For. Sci. Technol.* **2018**, *3*, 77–81. (In Chinese)
8. Li, Z.; Zhang, Y.; Wang, L. Phenolic resin modified by boron-silicon with high char yield. *Polym. Test.* **2019**, *73*, 208–213.
9. Wang, H.; Eberhardt, T.L.; Wang, C. demethylation of alkali lignin with halogen acids and its application to phenolic resins. *Polymers* **2019**, *11*, 1771. [CrossRef]
10. Yong, C.; Hou, X.; Wang, W. Synthesis and characterization of bio-oil phenol formaldehyde resin used to fabricate phenolic based materials. *Materials* **2017**, *10*, 668. [CrossRef]
11. Yu, Y.; Wang, Y.; Xu, P.; Chang, J. Preparation and characterization of phenolic foam modifified with bio-oil. *Materials* **2018**, *11*, 2228. [CrossRef]
12. Tiejun, G.; Xiapeng, T.; Hai, L. Research of phenolic resin and foam modified by melamine. *J. Shenyang Univ. Chem. Technol.* **2015**, *29*, 134–139.
13. Lu, C. Flame Retardant and Thermal degradation kinetics of PET/Aluminum di-ethyl phosphinate. *Plast. Sci. Technol.* **2016**, *44*, 26–29.
14. Yi, Z.; Zhang, J.; Zhang, S. Synthesis and mechanism of metal-mediated polymerization of phenolic resins. *Polymers* **2016**, *8*, 159. [CrossRef]
15. Chen, Y.; Mao, X.; Qian, L.; Xin, F.; Xu, G. Synergistic effect between aluminum hypophosphite and OMMT in flame-retarded PLA. *Eng. Plast. Appl.* **2015**, *41*, 19–24.
16. *ISO 4589-2 Plastics-Determination of Burning Behaviour by Oxygen Index–Part 2: Ambient-Temperature Test*; International Standards Organisation: Geneva, Switherland, 2017.
17. *ISO 5660 Fire Tests-Reaction to Fire-Part 1: Rate of Heat Release from Building Products (Cone Calorimeter Method)*; International Standards Organisation: Geneva, Switherland, 2015.
18. Tian, G.; Yuan, H. Determination of melamine concentration in milk powder and liquid milk by fourier transform-infrared spectroscopy. *Mod. Sci. Instrum.* **2009**, *5*, 86–91. (In Chinese)
19. Lu, Z.; Feng, W.; Kang, X. Research of flame retardant PA66 composites with benzoxazine and diethyl-phosphate aluminum. *Fine Chem.* **2019**, *2*, 322–327.

20. Shengnan, L.; Yangyang, Z.; Jingbo, Z. Synthesis of aluminum salts of diethyl phosphinic acid and its synergistic effect on flame retardant thermalplastic elastomer. *China Plast. Ind.* **2018**, *46*, 138–141.

21. Ding, Y.; Ezekoye, O.A.; Zhang, J. The effect of chemical reaction kinetic parameters on the bench-scale pyrolysis of lignocellulosic biomass. *Fuel* **2018**, *232*, 147–153. [CrossRef]

22. Zhou, X.; Li, J.; Wu, Y. Synergistic effect of aluminum hypophosphite and intumescent flame retardants in polylactide. *Polym. Adv. Technol.* **2015**, *26*, 255–265. [CrossRef]

23. Ding, Y.; Zhang, W.; Yu, L. The accuracy and efficiency of GA and PSO optimization schemes on estimating reaction kinetic parameters of biomass pyrolysis. *Energy* **2019**, *176*, 582–588. [CrossRef]

24. Xuan, Z.; Yang, L.; Danni, L. Preparation of Phenolic Resin Modified by Cardanol and Melamine. *Chin. J. Colloid Polym.* **2012**, *30*, 78–80.

25. Wang, B.; Qian, X.; Shi, Y. Cyclodextrin microencapsulated ammonium polyphosphate: Preparation and its performance on the thermal, flame retardancy and mechanical properties of ethylene vinyl acetate copolymer. *Compos. Part B Eng.* **2015**, *69*, 22–30. [CrossRef]

26. Braun, U.; Horst, B.; Heinz, S. Flameretardancy mechanisms of metal phosphinates and metal phosphinates in combination with melamine cyanurate in glass-fiber reinforced poly(1,4-butyleneterephthalate): The influence of metal cation. *Polym. Adv. Technol.* **2008**, *19*, 680–692. [CrossRef]

27. Braun, U.; Schaael, B. Flameretardancy mechanisms of aluminum phosphinate in combination with melamine cyanuratein glass-fiber reinforced poly(1,4 -butylene terephthalate). *Macromol. Mater. Eng.* **2008**, *293*, 206–217. [CrossRef]

28. Chen, X.; Song, W.; Liu, J. Synergistic flame-retardant effects between aluminum hypophosphite and expandable graphite in silicone rubber composites. *J. Therm. Anal. Calorim.* **2015**, *120*, 1819–1826. [CrossRef]

29. Ding, Y.; Fukumoto, K.; Ezekoye, O.A. Experimental and numerical simulation of multi-component combustion of typical charring material. *Combust. Flame* **2020**, *211*, 417–429. [CrossRef]

30. Zhan, Z.; Xu, M.; Li, B. Synergistic effect of nano-silica on aluminum Diethyl phthalate/polyamide 66 system. *2015 China Flame Retard. Acad. Annu. Conf. Pap.* **2015**, 383–392. [CrossRef]

31. Xu, S.; Qi, L.; Ma, C. Preparation and properties of modified melamine resin blended viscose fiber. *Wool Text. J.* **2018**, *46*, 11–14.

 *materials*

Article

Investigation on the Flame Retardant Properties and Fracture Toughness of DOPO and Nano-SiO$_2$ Modified Epoxy Novolac Resin and Evaluation of Its Combinational Effects

Markus Häublein [1], Karin Peter [2], Gökhan Bakis [3], Roi Mäkimieni [2], Volker Altstädt [1],* and Martin Möller [4],*

[1] Department Polymer Engineering, University of Bayreuth, Universitätsstraße 30, 95447 Bayreuth, Germany; markus.haeublein@uni-bayreuth.de

[2] DWI–Leibniz-Institute for Interaktive Materials Aachen, Forckenbeckstraße 50, 52056 Aachen, Germany; peter@dwi.rwth-aachen.de (K.P.); maekimieni@dwi.rwth-aachen.de (R.M.)

[3] BASF Polyurethanes GmbH, Elastogranstraße 60, 49448 Lemfoerde, Germany; goekhan.bakis@basf.com

[4] ITMC-Institute of Technical and Macromolecular Chemistry, RWTH Aachen University, Worringer Weg 1, 52074 Aachen, Germany

* Correspondence: altstaedt@uni-bayreuth.de (V.A.); moeller@dwi.rwth-aachen.de (M.M.)

Received: 11 April 2019; Accepted: 8 May 2019; Published: 10 May 2019

Abstract: In this study, the flame-retardant, thermal and mechanical properties of 9,10-dihydro-9-oxa-10-phosphaphenanthrene-10-oxide (DOPO) and nano-SiO$_2$ modified epoxy novolac resin is evaluated, and the combinational effects of both additives are verified. As a hardener, an isophorone diamine (IPDA) and polyetheramine blend is stoichiometrically added to obtain a low viscous epoxy resin system, suitable for resin injection and infusion techniques. The glass transition temperature (Tg) and the silica dispersion quality is affected by the DOPO modification and the nano silica particles. The flame-retardant (FR) and mechanical properties of the additives are investigated separately. The fracture toughness could be increased with the incorporation of both FR additives; however, the effect is deteriorated for higher DOPO amount which is referred to silica particle agglomeration and consequently reduced shear yielding mechanism. Flame-retardant properties, especially the peak heat release rate (pHRR) and the total heat release (THR) could be decreased from 1373.0 kW/m^2 of neat novolac to 646.6 kW/m^2 measured by resins with varying phosphorous and silica content. Thermogravimetric analysis (TGA) measurements show the formation of a high temperature stable char layer above 800 °C which is attributed to both additives. Scanning electron microscopy (SEM) images are taken to get deeper information of the flame-retardant mechanism, showing a dense and stable char layer for a certain DOPO silica mixture which restrains the combustible gases from the burning zone in the cone calorimeter test and influences the fire behavior of the epoxy resin.

Keywords: epoxy novolac resin; DOPO; nano-SiO$_2$; flame retardancy; fracture toughness

---

1. Introduction

Since the first patent regarding epoxy and amine systems was submitted in 1934 by Paul Schlack in Germany the importance of resin has gathered increasing relevance. There are several reasons for the wide application range of epoxys nowadays, for instance their good processability with low cure shrinkage and therefore lower residual stresses in the cured part, good mechanical properties and outstanding customization which is verified to the variation of properties with different hardener and resin selection [1]. However, there are also several disadvantages, for example hot-wet and flame-retardant properties, compared to other thermosets, like bismaleimides or phenolic resins which show good intrinsic flame-retardant properties [1].

To overcome the lower flame resistance, different additives like halogenated [2,3] or phosphorous-based [4,5] FRs as well as inorganic fillers (like aluminum trihydroxide ATH) [6] are incorporated into the epoxy matrix and the properties evaluated in the past decades. However, halogenated flame retardants release toxic gases during decomposition and are therefore prohibited by the REACH (Registration, Evaluation, Authorization and Restriction of Chemicals) regulation [7]. Otherwise, inorganic fillers require a high content [6] for the improvement of fire resistance which worsens the processing of the resin system and downgrades its mechanical properties. For these reasons, research in the last decade focuses on phosphorous based epoxy resin systems, especially the 9,10-dihydro-9-oxa-10-phosphaphenanthrene-10-oxide (DOPO) molecule indicates promising flame-retardant properties.

Wang et al. [4] studied the effect of DOPO modified phenolic amine hardeners on the flame-retardant properties of diglycidyl ether of bisphenol A (DGEBA, glass transition temperature, $T_g = 171.8$ °C) based epoxy resin, showing that 0.83 wt.% phosphorous in the overall system sufficient to reach the UL94-V0 rating and to increase the limiting oxygen index (LOI) from 25.3% for DGEBA up to 36.1% for a modified system. Another study by Wang et al. [5] investigated the influence of DOPO modification in the resin backbone. Tetradiglycidyl diaminodiphenyl methane (TGDDM, $T_g = 249$ °C) was modified with DOPO and as a hardener 4,4′-diaminodiphenyl sulfone (DDS) was added. The UL94-V0 rating could be reached with only 0.8 wt.% phosphorous and the LOI is increased from 29.1% up to 31.8% for the modified system.

DOPO modified epoxy resin as one component of a flame retarding synergistic system consisting of silica as second component, was investigated by Zhang et al. in 2012 [8]. A DGEBA-m-Phenylendiamine (m-PDA) composite containing oligomeric silsesquioxane (OPS) and DOPO was prepared, while the overall FR-content was kept constant at 5 wt.%. The lowest LOI was found for the composite with 1.25 wt.% OPS and 3.75 wt.% DOPO (LOI = 30.3%), the UL94-V1 rating was verified for every combined system, whereas the lowest peak heat release rate (p-HRR) was determined for 2.5 wt.% OPS and DOPO (p-HRR = 603 kW/m^2) content. The synergistic effect of the system was verified for a DOPO and silica modified system and justified with the blowing out effect.

However, the synergistic effect of phosphorus-based FR and nano-SiO$_2$ particles as a silicon containing additive and the effectiveness in a low T_g-epoxy resins as well as the influence of the flame retardants on the mechanical properties of the epoxy resin system has not been investigated yet. In this study, the fracture toughness, decomposition behavior and flame-retardant properties (Thermogravimetric analysis: TGA and cone calorimeter) are studied for a nano-SiO$_2$ and DOPO containing epoxy novolac resin system. In addition, the synergistic effect is discussed, and the structure-property relationship established with scanning and transmission electron microscopy images.

2. Materials and Methods

2.1. Materials

Epoxy novolac resin D.E.N. 431 (epoxy equivalent weight 176 g/mol) is purchased from Olin. To reduce overall viscosity of the novolac resin 5 wt.% (based on the D.E.N. 431 content) of the reactive diluent Heloxy modifier BD (Hexion, Columbus, OH, USA, epoxy equivalent weight 124 g/mol) is added. As a curing agent isophorone diamine (IPDA, Aradur 42, Huntsman, TX, USA, amino equivalent weight 42 g/mol) and the polyetheramine D230 (Jeffamine XB3403, Huntsman, TX, USA, amino equivalent weight 60 g/mol) in the ratio 50:50 by weight were used. As a flame retardant (FR), 9,10-dihydro-9-oxa-10-phosphaphenanthrene-10-oxide (DOPO) modified epoxy novolac (Struktol VP 3760, Schill&Seilacher, Hamburg, Germany, epoxy equivalent weight 340 g/mol) and spherically shaped nano-SiO$_2$-particles as an epoxy novolac master batch (NANOPOX F700, Evonic Industries,

Essen, Germany, epoxy equivalent weight 310 g/mol) were added in varying amounts. The chemical structure of the different materials is shown in Figure 1.

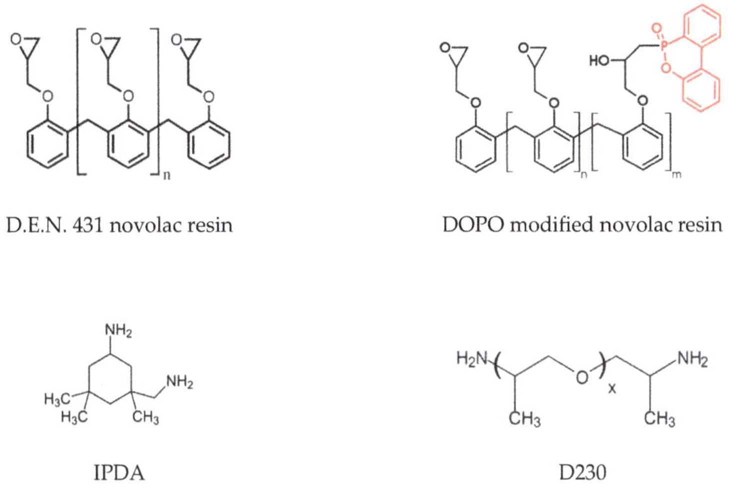

D.E.N. 431 novolac resin

DOPO modified novolac resin

IPDA

D230

Figure 1. Chemical structure of the epoxy novolac resin system (DOPO modification is highlighted red) and the hardener (blend ratio 50:50 parts by weight).

2.2. Methods

The epoxy novolac resin are weighted and mixed at room temperature with 1700 rpm using a DAC 150.1 FVZ speedmixer. The mixture is degassed in a vacuum chamber. Different amount of FR is generated by replacing some of the D.E.N. 431 by the DOPO- and/or nano-SiO_2-modified novolac master batches. For the mixture with 1 wt.%, 2 wt.% and 3 wt.% phosphorous (Novolac + 1 wt.% P, Novolac + 2 wt.% P and Novolac + 3 wt.% P), 1/6, 1/3 and 1/2 of the resin is replaced by the DOPO-master batch, respectively. After the stoichiometric amount of the hardener blend is added, the mixture is homogenized with the speedmixer and degassed again, to achieve neat resin plates without air inclusions. The liquid resins system is moulded in a steel plate, which is pre-treated with the release agent Loctite Frekote 770NC to ensure demould, and fully cured at 80 °C for 60 min and 140 °C for 90 minutes (heating rate: 5 K/min) in an oven.

Dynamic mechanical thermo analysis (DMTA) is performed according to DIN EN ISO 6721-2 with a Rheometric Scientific RDA III, the frequency is set to 1 Hz and the deformation to 0.1%. Glass transition temperature is determined by storage modulus (G') onset, two samples are tested for each material.

The fracture toughness of the material systems is evaluated by the critical stress intensity factor in mode I (K_{Ic}-value) according to ISO 13586 with a universal testing machine Zwick Z050 (Zwick Roell, Ulm, Germany). The initial crack is performed by tapping a razor blade into the notch. The initial load is set to 2 N and the testing speed to 10 mm/min, five specimens are tested. The critical energy release rate in mode I loading (G_{Ic}-value) is calculated using the Young's modulus (E) and the following Equation (1) [9]. The modulus was determined with tensile tests of the materials.

$$G_{Ic} = K_{Ic}^{2}/E,\qquad(1)$$

TGA measurements using a Netzsch 209 F1 libra (Netzsch, Selb, Germany) are performed under air conditions to evaluate the degradation behavior and the silica content of the material systems. Samples with the weight of 10 to 25 mg are used and heated up from room temperature to 1000 °C with 10 K/min.

Flame retardant properties are investigated with a cone calorimeter iCone (Fire Testing technology FTT, East Grinstead, UK) with the heat flux of 50 kW/m^2 and a distance between heater and sample of 60 mm. Three samples are tested for each material. The discussed values were averaged, and the standard deviation calculated, one curve is shown representatively in the graphs.

For Raman measurements, a WITec alpha 300 RA+ system equipment with a UHTS 300 spectrometer and a back-illuminated Andor Newton 970 EMCCD camera (Oxford Instruments, Abingdon, UK) was used. The measurements were conducted employing an excitation wavelength of $\lambda = 532$ nm and the spectra were recorded with an integration time of 1 s and 50 accumulations (50 x LWD objective, NA = 0.55). All spectra were subjected to a cosmic ray removal routine and baseline correction (software WITec Project FIVE, version 5.1, WITec, Ulm, Germany).

To study the toughening mechanism and the flame-retardant effect, scanning electron microscopy (SEM) using a Zeiss Leo 1530 instrument (Carl Zeiss, Oberkochen, Deutschland) at an acceleration voltage of 3 kV and transmission electron microscopy (TEM) using a Zeiss EM 922 Omega (Carl Zeiss, Oberkochen, Deutschland) with 200 kV acceleration voltage are performed. For SEM studies the sample is coated with a Cressington Sputter Coater to create a 1.3 nm thick conductive platinum layer; images were taken with the in lens and the secondary electron detector. Sample preparation for TEM analysis are performed with an ultra-microtome (Leica EM UC 7, Leica Camera, Wetzlar, Germany) at room temperature. The resulting 50 nm thick samples are placed on a carbon coated copper grid.

3. Results and Discussion

3.1. Effect of DOPO on Epoxy Resin Properties

Glass transition temperature and fracture toughness of DOPO modified epoxy resin systems can be seen in Table 1. The glass transition temperature shows decreasing values for the DOPO-modified system as the Tg decreases from 109.8 °C for neat novolac to 66.5 °C for 3 wt.% phosphorous containing resin. This change is given by the chemical structure of the DOPO-modified system; the DOPO molecule is chemically bound to the epoxy group of the novolac (see Figure 1) which is consequently not involved in crosslinking and network formation of the cured resin system for that reason [10,11]. The fracture toughness of the systems shows a contrary trend as it seems to be increasing for higher amount of DOPO. Neat novolac reference shows a G_{Ic}-value of 109.2 J/m^2, whereas the highest value (131.8 J/m^2) is measured with 3 wt.% of phosphorous, but considering standard deviation, they are in the same range.

Table 1. Adjusted phosphorous content, glass transition temperature and fracture toughness of DOPO modified epoxy resin.

	Phosphorous Content/wt.%	Tg/°C	G_{Ic}/J/m^2
Novolac (Reference)	0.0	109.8 ± 0.3	109.2 ± 24.4
Novolac + 1 wt % P	1.0	96.7 ± 0.5	123.8 ± 18.6
Novolac + 2 wt % P	2.0	81.9 ± 0.9	107.3 ± 12.0
Novolac + 3 wt % P	3.0	66.5 ± 0.6	131.8 ± 27.5

The flame-retardant properties of the DOPO modified epoxy resin systems were evaluated with TGA under air conditions and cone calorimeter experiments to study the decomposition and burning behavior. The mass loss rate from the TGA and the heat release rate from the cone calorimeter measurement of the modified systems is shown in Figure 2, a summary of the relevant parameters from TGA and cone is given in Table 2. It is represented in the TGA and cone calorimeter curves that the DOPO additive improve the flame retardancy of the resin system. First, it enhances the char yield at higher temperature, which can be seen in Figure 2 on the left side and in Table 2 since it increases with higher phosphorous content (residual mass @700 °C and @900 °C, see Table 2). The energy release (peak heat release, pHRR and total heat release, THR) is also improved with increasing DOPO content, for highest phosphorous amount, the pHRR and the THR can be reduced from 1373.0 kW/m^2

and 87.7 MJ/m^2 for neat novolac to 689.2 kW/m^2 and 57.4 MJ/m^2. The reduction in heat release is referred to the gas phase since phosphorous containing flame retardants decompose and release phosphorous-oxide radicals to reduce overall heat release. However, the systems start to decompose earlier for higher phosphorous content, as the T$_{d5\%}$-value (the temperature when 5% of the systems is degraded) as well as the time to ignition (tti) are reduced. It is estimated that the resin absorbs energy and starts to decompose earlier, as the glass transition temperature and network density is reduced with a higher amount of phosphorous.

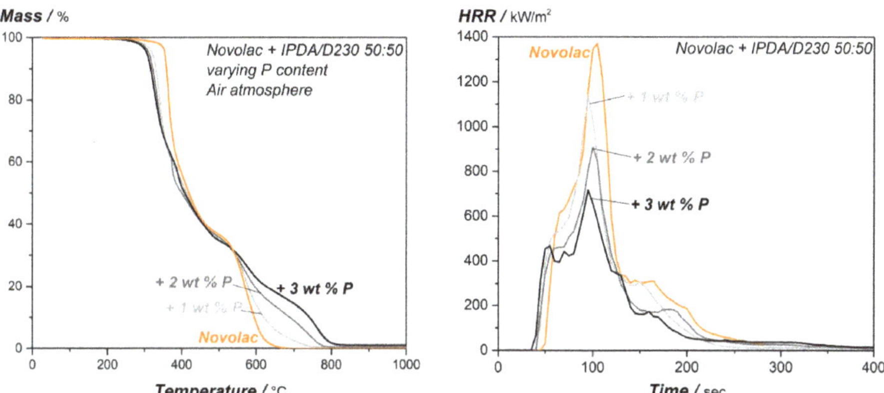

Figure 2. TGA measurement under air atmosphere (**left**) and cone calorimeter test (**right**; 50 kW/m^2 and distance sample to heater: 60 mm) of the DOPO modified epoxy resin.

Table 2. T$_{d5\%}$, residual mass @700 °C and @900 °C from TGA under air atmosphere and time to ignition (tti), peak heat release rate (pHRR) and total heat release (THR) from cone calorimeter experiments for DOPO modified epoxy resins.

	TGA			Cone Calorimeter		
		Residual Mass				
	T$_{d5\%}$/°C	@700 °C/wt.%	@900 °C/wt.%	tti/sec	pHRR/kW/m^2	THR/MJ/m^2
Novolac (Reference)	354.6	0.1	0.1	53.7 ± 2.9	1373.0 ± 151.6	87.7 ± 1.3
Novolac + 1 wt.% P	320.5	2.5	0.4	45.7 ± 1.5	1131.3 ± 59.8	73.3 ± 4.2
Novolac + 2 wt.% P	311.7	8.9	0.4	42.7 ± 0.6	917.7 ± 43.4	64.6 ± 1.5
Novolac + 3 wt.% P	306.2	14.4	1.0	41.0 ± 2.0	689.2 ± 43.2	57.4 ± 1.1

3.2. Effect of Nano-SiO$_2$ on Epoxy Resin Properties

The influence of silica particles on glass transition and fracture toughness is summarized in Table 3. It is shown that the Tg is increased for the nano-SiO$_2$ modified novolac resin which is referred to the surface modification of the silica particles. It is expected that the surface modified particles show high interaction with the epoxy resin which results in an ordered structure of the epoxy resin matrix around the particles. The ordered particle-matrix interphase needs higher thermal energy to start the polymer chain vibration. The compatibility of the commercial NANOPOX product with epoxy resin systems is already discussed in the literature and demonstrated by increasing Tg [12]. In addition, the fracture toughness is also improved for higher silica content from 109.2 J/m^2 for neat novolac to 207.6 J/m^2 for 3.9 vol.% SiO$_2$ particles. The toughening mechanism for nano silica particles is discussed in the literature already and basically referred to the shear yielding mechanism, a deeper evaluation of toughening effect for the systems given here will be discussed in Section 3.3.

Table 3. Adjusted silica content, glass transition temperature and fracture toughness of nano-SiO$_2$ modified epoxy resin.

	Nano-SiO$_2$ Content/vol.%	Tg/°C	G$_{Ic}$/J/m^2
Novolac	0.0	109.8 ± 0.3	109.2 ± 24.4
Novolac + 0.8 SiO$_2$	0.8	111.3 ± 1.3	141.5 ± 34.6
Novolac + 1.5 SiO$_2$	1.5	111.0 ± 0.5	155.5 ± 16.1
Novolac + 2.3 SiO$_2$	2.3	110.1 ± 0.1	191.9 ± 33.2
Novolac + 3.9 SiO$_2$	3.9	112.5 ± 0.4	207.6 ± 16.5

TEM images are taken to evaluate the dispersion quality of the silica nano-particles and to ensure that cured epoxy resins only show slight amount of small agglomerates. The TEM images are represented in Figure 3 for lowest and highest nano-SiO$_2$ content and 2.3 vol.% SiO$_2$. It can be summarized that the particles are homogeneously distributed, and no agglomerates are formed even at higher content.

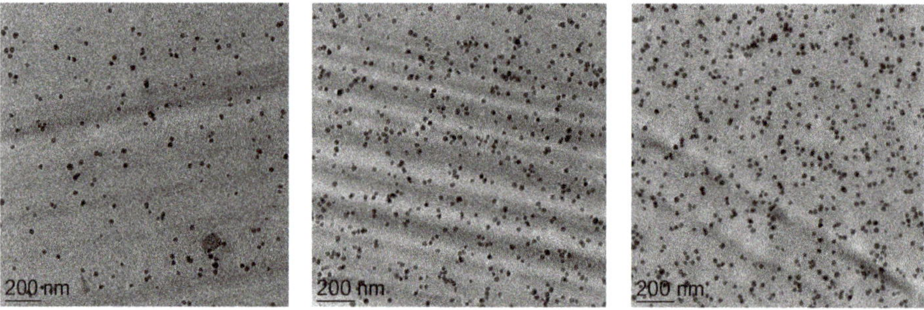

Figure 3. TEM images of nano-SiO$_2$ modifed epoxy resin: 0.8 vol.% (**left**), 2.3 vol.% (**middle**) and 3.9 vol.% (**right**).

Decomposition and burning behavior of nano-silica modified epoxy resin is shown in Figure 4 and summarized in Table 4. The decomposition behavior on the left side of Figure 4 is only influenced by the silica particles, the residual mass at higher temperature is basically represented by the adjusted silica amount, the T$_{d5\%}$-value is reduced for the silica nanocomposites to 338.5 °C and 343.2 °C as highest and lowest value respectively, however the effect is only marginal. Silica amount shows only slight impact on the heat release values and time to ignition from the cone calorimeter experiments. Whereas the peak heat release rate is continuously decreasing from 1373.0 kW/m^2 for the neat resin to 910.9 kW/m^2 for a resin containing 3.9 vol.% SiO$_2$, the total heat release is decreased for highest silica content. However, the THR increases for lower nano SiO$_2$ amount which can be explained by a supportive effect of the particles as they raise the effective surface for the flame. The reduction in THR and especially pHRR can be explained by the amount of non-flammable silica particles. The time to ignition is unaffected for the tested nanocomposites considering standard deviation. It can be summarized that nano-SiO$_2$ shows only a slight impact on the flame-retardant properties of the epoxy resin, especially for lower particle loading, compared to the DOPO modified systems for instance.

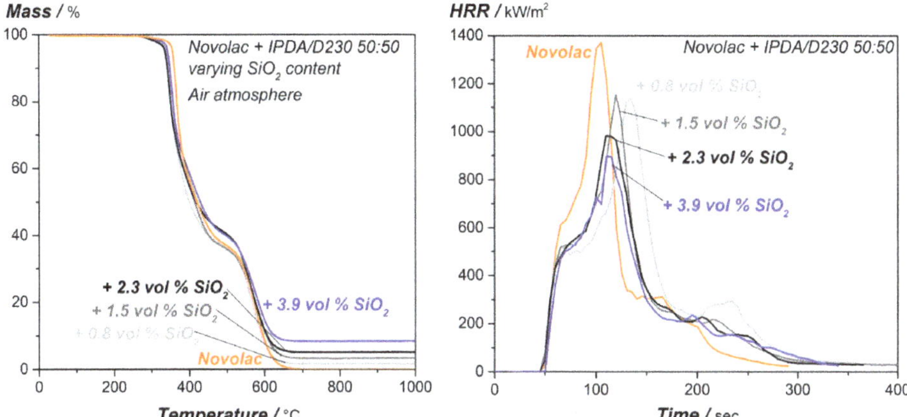

Figure 4. TGA measurement under air atmosphere (**left**) and cone calorimeter test (**right**; 50 kW/m^2 and distance sample to heater: 60 mm) of the silica modified epoxy resin.

Table 4. $T_{d5\%}$, residual mass @700 °C and @900 °C from TGA under air atmosphere and time to ignition (tti), peak heat release rate (pHRR) and total heat release (THR) from cone calorimeter experiments for nano-SiO$_2$ modified epoxy resins.

		TGA		Cone Calorimeter		
	$T_{d5\%}$/°C	Residual Mass		tti/sec	pHRR/kW/m^2	THR/MJ/m^2
		@700 °C/wt.%	@900 °C/wt.%			
Novolac	354.6	0.1	0.1	53.7 ± 2.9	1373.0 ± 151.6	87.7 ± 1.3
Novolac + 0.8 SiO$_2$	339.2	1.7	1.5	53.3 ± 1.2	1138.9 ± 53.3	103.9 ± 3.3
Novolac + 1.5 SiO$_2$	338.6	3.3	3.3	52.7 ± 1.5	1141.9 ± 84.4	97.0 ± 2.0
Novolac + 2.3 SiO$_2$	338.5	5.1	5.0	52.7 ± 1.2	1022.4 ± 34.7	90.2 ± 1.0
Novolac + 3.9 SiO$_2$	343.2	8.6	8.6	54.7 ± 0.6	910.9 ± 58.3	84.4 ± 2.9

The residual cone calorimeter char layer was investigated with Raman spectroscopy to evaluate the flame-retardant effect of the silicon-oxide particles. The optical microscopy images (10× and 50× magnification) taken at the burning zone and the Raman spectra of the sample with 2.3 vol.% SiO$_2$ are presented in Figure 5. The orange cross indicates the position, where the Raman spectrum was recorded. The microscopy images reveal that the nano-SiO$_2$ particles generate a white appearing, dense and closed glass layer, which results during the burning process of the matrix material. The layer is composed of silicon oxide basically, since the Raman spectrum shows the typical SiO$_2$ peaks discussed in the literature [13]. In contrast to this, the black Raman spectrum, which was recorded at the inside (the opposite side to the burning zone) of the cone calorimeter sample, indicates the peaks resulting from carbon black, according to Pawlyta et al. [14]. The different Raman spectra from the inside and the outside (burning zone side) of the nano-SiO$_2$ modified sample show that silicon oxide accumulates at the burning side of the sample and generates a dense and closed layer.

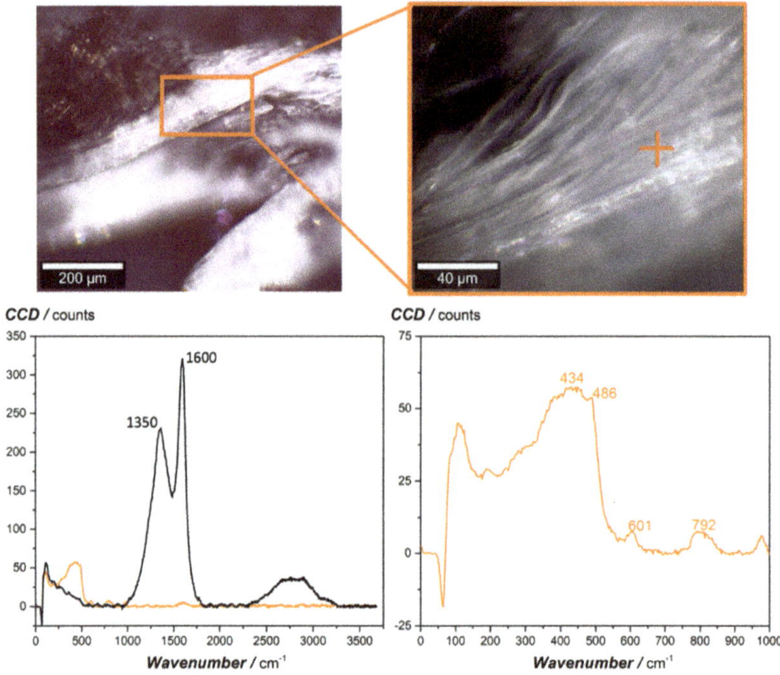

Figure 5. Optical microscopy images of the residual cone calorimeter char of the sample with 2.3 vol.% SiO$_2$ with lower and higher magnification and the corresponding Raman spectrum taken at the dense, white appearing layer on top of the burning zone (orange spectrum, measurement position indicated by orange cross in higher magnified microscopy image). The black Raman spectrum was recorded from the inside (opposite side of the burning zone) of the cone colorimeter sample.

3.3. Additive Combination: Effect on Glass Transition Temperature, Dispersion Quality and Fracture Toughness Modelling of Modified Systems

So far, both additives are investigated and evaluated separately; next, the influence on the properties of the combination of both additives will be discussed. The impact on glass transition temperature and fracture toughness is represented in Table 5 and compared to the appropriate system without silica particles. The combination shows a reduction in Tg for 2 wt.% and 3 wt.% phosphorous with higher silica amount. For 2 wt.% and 3 wt.% phosphorous, the glass transition is reduced from 81.9 °C to 75.6 °C by adding 4.4 vol.% SiO$_2$ for 2 wt.% P and from 66.5 °C to 62.2 °C for 3 wt.% phosphorous and 4.7 vole% silica. It is represented in the TEM microscopy images in Figure 6 that the dispersion quality of the nano-silica particles is deteriorated for material systems with higher DOPO amount as there are some particle agglomerates visible in the micrographs. Due to agglomeration of silica particles, the particle-matrix interphase is weakened, resulting in continuously decreasing T$_g$ for higher SiO$_2$-content. The reason for lower dispersion quality is estimated to occur due to the shift of the chemical potential of the liquid resin due to the DOPO modified novolac which is not optimized for the commercial silica particle modification. The particle-particle interaction is estimated to be higher leading to increased agglomerate number.

Table 5. Adjusted silica content, glass transition temperature and fracture toughness of DOPO and nano-SiO$_2$ modified epoxy resin.

	Phosphorous Content/wt.%	Nano-SiO$_2$ Content/vol.%	Tg/°C	G$_{Ic}$/J/m^2
Novolac + 2 P	2.0	0.0	81.9 ± 0.9	107.3 ± 12.0
Novolac + 2 P + 2.7 SiO$_2$	2.0	2.7	78.2 ± 0.1	201.9 ± 31.4
Novolac + 2 P + 4.4 SiO$_2$	2.0	4.4	75.6 ± 0.1	220.8 ± 15.9
Novolac + 3 P	3.0	0.0	66.5 ± 0.6	131.8 ± 27.5
Novolac + 3 P + 2.8 SiO$_2$	3.0	2.8	63.8 ± 0.1	206.8 ± 37.3
Novolac + 3 P + 4.7 SiO$_2$	3.0	4.7	62.2 ± 0.1	243.1 ± 17.4

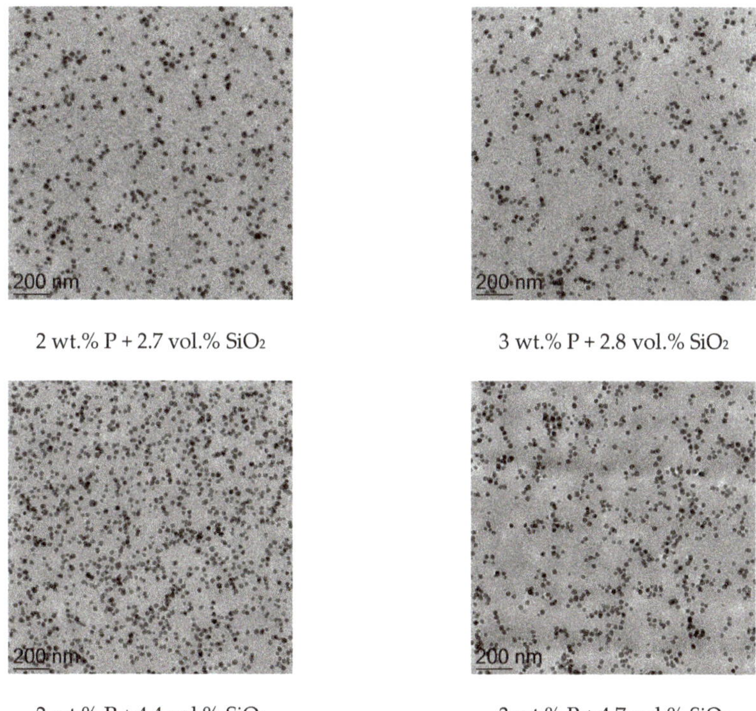

2 wt.% P + 2.7 vol.% SiO$_2$

3 wt.% P + 2.8 vol.% SiO$_2$

2 wt.% P + 4.4 vol.% SiO$_2$

3 wt.% P + 4.7 vol.% SiO$_2$

Figure 6. Transmission electron microscopy images of composites with 2 wt.% (**left**) and 3 wt.% (**right**) phosphorous and comparable silica content.

The G$_{Ic}$-value from Table 5 for the combined systems is plotted in Figure 6 for varying silica content. It can be seen that the nano silica particles show high impact on fracture toughness of modified epoxy resin, increasing from 107.3 J/m^2 and 131.8 J/m^2 for 2 wt.% and 3 wt.% P up to 220.8 J/m^2 (2 wt.% P) and 243.1 J/m^2 (3 wt.% P) for 4.4 vol.% and 4.7 vol.% SiO$_2$ respectively. In addition, the dotted lines in Figure 7 represent a model of the toughening effect of the shear yielding mechanism for nano-sized particles in epoxy resin systems proposed by Hsieh et al. and Johnsen et. al. [15,16]. A particle size of 20 nm is assumed for the calculation according to the datasheet of the supplier of silica particles. It is shown that the modelling fits to the measured values for 0 wt.% phosphorous and the 2 wt.% phosphorous epoxy resin system. This confirms that the dominating toughening mechanisms is shear yielding of 20 nm sized SiO$_2$-particles. However, the toughening contribution from shear yielding induced by nano-SiO$_2$ particles is overestimated with the modelling for the epoxy resin with 3 wt.% phosphorous. As it is shown in the TEM micrographs in Figure 6, the DOPO modification seems to support particle agglomeration. The overall average particle size is increased for this reason and consequently the shear yielding mechanisms is deteriorated as higher particle sizes lead to lower fracture toughness for this model.

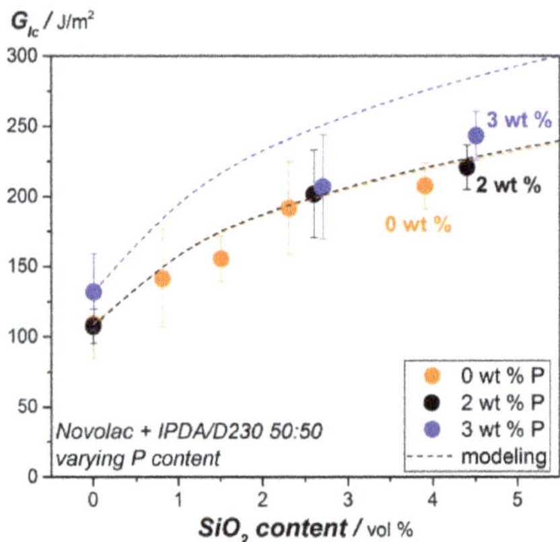

Figure 7. Fracture toughness of the tested material systems.

To get deeper information of the fracture and toughening mechanism, the fracture surface is studied with SEM (see Figure 7). For comparison, the fracture surface of neat novolac resin and the fracture surface of only silica modified epoxy resin with comparable nano-SiO_2 content is represented. For the neat novolac reference a flat and smooth fracture surface is developing during crack propagation which indicates the brittle character of the unmodified system and the low G_{Ic}-value of only 109.2 J/m². However, there are also some river lines visible on the fracture surface of the epoxy resin, indicating plastic deformation and shear yielding mechanism.

For rigid nano silica epoxy resin, different toughening mechanism are already discussed in the literature [17–19], these are, in general: shear yielding and particle pull-out and plastic void growth of the matrix [17–19]. In Figure 8, for the only silica modified system, the different toughening effects are demonstrated on the fracture surface. However, the particle pull-out toughening mechanism could hardly be verified for the presented systems. Shear yielding seems to be the dominating toughening mechanisms for the system presented here. The river like structure on the fracture surface of silica modified systems in Figure 8 is reflected to the shear yielding mechanism and plastic deformation of the matrix which is more dominant and effective for the SiO_2 modified epoxy resin compared to neat novolac resin. For the DOPO modified epoxy resins, the river like structure is still pronounced on the surface, especially for the 2 wt.% phosphorous. However, the tendency of silica particles to agglomerate for higher phosphorous content seems to reduce plastic deformation and shear yielding mechanism as the river like structure is less dominant compared to lower phosphorous content. The deteriorating effect has already been discussed previously for the modelling studies (see Figure 7).

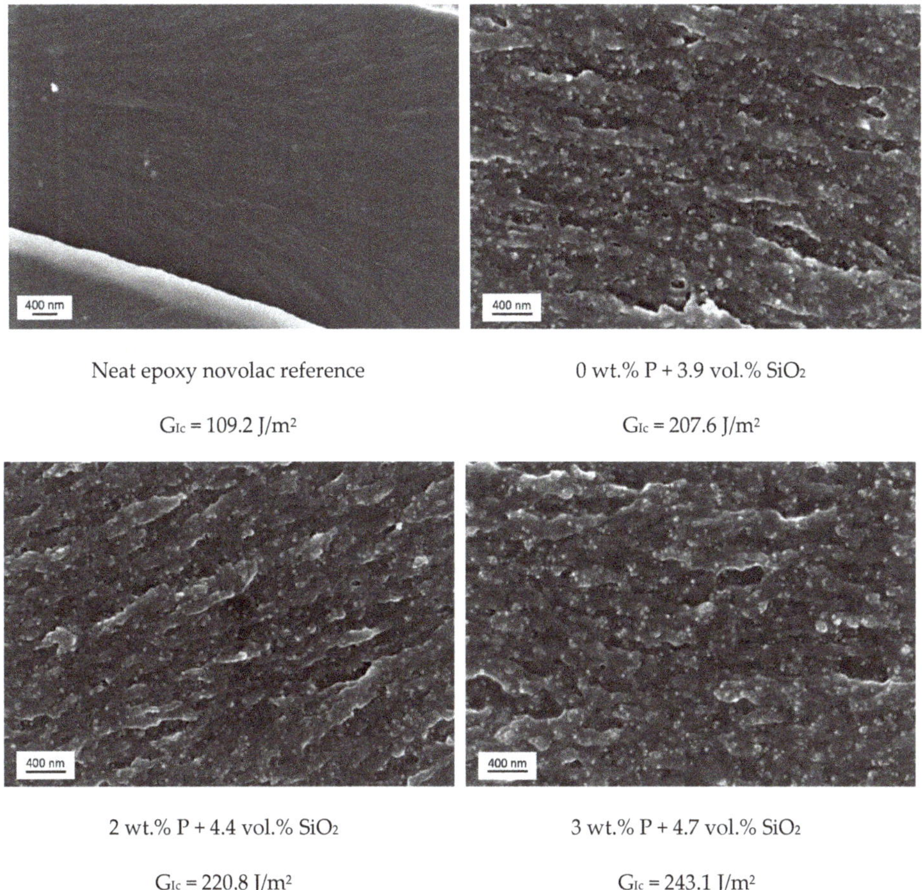

Neat epoxy novolac reference

G_{Ic} = 109.2 J/m^2

0 wt.% P + 3.9 vol.% SiO$_2$

G_{Ic} = 207.6 J/m^2

2 wt.% P + 4.4 vol.% SiO$_2$

G_{Ic} = 220.8 J/m^2

3 wt.% P + 4.7 vol.% SiO$_2$

G_{Ic} = 243.1 J/m^2

Figure 8. SEM images of K_{Ic}-fracture surface of neat novolac (up left), novolac + 3.9 vol.% SiO$_2$ (up right) novolac + 2 wt.% phosphorous + 4.4 vol.% SiO$_2$ (down left) and novolac + 3 wt.% phosphorous + 4.7 vol.% SiO$_2$ (down right). Crack propagation direction: left to right.

3.4. Additive Combination: Flame Retardancy of DOPO and Silica Modified Novolac Resin

To evaluate the degradation behavior and flame retarding effects of the DOPO and silica modified system, TGA (air atmosphere) and cone calorimeter measurements are performed. The resulting TGA mass loss curves for the reference system without DOPO, with 2 wt.% and 3 wt.% phosphorous and varying silica content are given in Figure 9, the summary ($T_{d5\%}$, residual mass @700 °C and @900 °C) is given in Table 6.

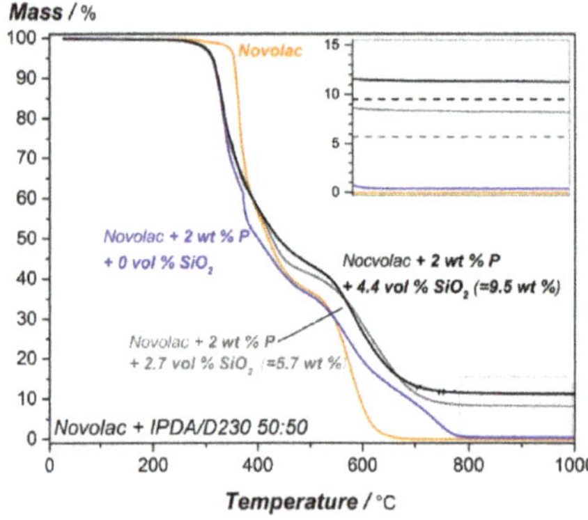

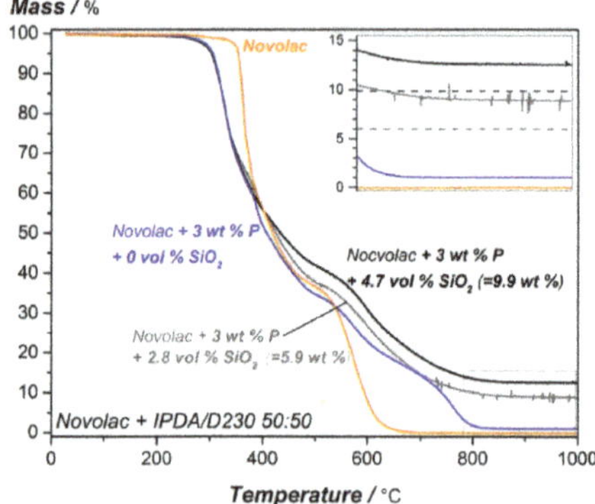

Figure 9. TGA mass loss curve for the modified epoxy novolac resin systems. Influence of silica content on 2 wt.% (**left**) and 3 wt.% (**right**) phosphorous modified resin.

Table 6. $T_{d5\%}$, residual mass @700 °C and @900 °C from TGA curves and deviation according to Equation (2).

	Nano-SiO$_2$ Content: m_{SiO2}/wt.%	$T_{d5\%}$/°C	Residual Mass @700 °C: $m_{700°C}$/wt.%	Residual Mass @900 °C: $m_{900°C}$/wt.%	Deviation @900 °C: $\Delta m_{900°C}$/%
Novolac (Reference)	0.0	353.9	0.1	0.0	0.0
Novolac + 2 P	0.0	311.7	8.9	0.4	0.0
Novolac + 2 P + 2.7 SiO$_2$	5.7	314.0	11.5	8.3	36.1
Novolac + 2 P + 4.4 SiO$_2$	9.5	313.0	13.3	11.3	14.1
Novolac + 3 P	0.0	306.2	14.4	1.0	0.0
Novolac + 3 P + 2.8 SiO$_2$	5.9	303.1	14.7	9.0	30.4
Novolac + 3 P + 4.7 SiO$_2$	9.9	303.1	18.9	12.6	15.6

The silica content does not show any influence on the decomposition behavior of the novolac epoxy resin at lower temperatures for the DOPO modified systems. The $T_{d5\%}$ is nearly unaffected, compared to the appropriate only DOPO modified system. Compared to this, the residual mass at higher temperature (900 °C) is affected by the combination of DOPO and silica. To quantify this in detail, the deviation in between the adjusted silica content and the measured residual mass of the systems at 900 °C ($\Delta m_{900°C}$) is calculated with the following equation.

$$\Delta m_{900°C} = (m_{DOPO+SiO2\,900°C} - m_{DOPO\,900°C} - m_{SiO2})/(m_{DOPO\,900°C} + m_{SiO2}), \tag{2}$$

Equation (2) includes the char yield of the appropriate only DOPO modified epoxy resin at 900 °C ($m_{DOPO\,900°C}$ from Table 6), the adjusted silica content (m_{SiO2} from Table 6) and the measured residual mass of the nano-silica and phosphorous modified novolac resin ($m_{DOPO+SiO2\,900°C}$: char yield at 900 °C for DOPO and silica modified system from Table 6).

The deviation is graphically represented in Figure 9 as mismatch between the measured TGA-curve and the adjusted SiO$_2$ content (dotted lines) for the combined systems. If the flame retardants would not interact with each other, the deviation should be 0.0% which means that the residual mass is equal to the adjusted silica content. However, as there is a higher deviation observed, it is estimated that a residual char product is generated being composed of both flame-retardant additive types (phosphorous and silica) having high temperature stability as it is stable in the measured area up to 1000 °C. The high temperature stability of the residual product also shows that it is highly oxidized as there is no further degradation and mass loss in air atmosphere. The deviation is in the range of 14–15% for higher and up to 30% and 36% for lower silica content, the biggest difference is given for the 2 wt.% phosphorous and 2.7 vol.% SiO$_2$ modified system.

To get deeper information about the flammability and decomposition behavior of the material, cone calorimeter tests are performed. The resulting HRR curves are shown in Figure 10 and summarized in Table 7. It is shown that the combination of additives continuously increases the pHRR for the epoxy novolac resin with 3 wt.% phosphorous from 689.2 kW/m^2 to 814.2 kW/m^2, further the total heat release is not improved for these systems. However, the time to ignition is nearly unaffected by the combination of DOPO and silica. For the novolac resin with 2 wt.% phosphorous and different silica content, the p-HRR shows lowest value (646.6 kW/m^2) for 2 wt.% phosphorous and 2.7 vol.% SiO$_2$. For higher silica content, the HRR increases again, this tendency is also reflected for the THR. The time to ignition is in addition elevated for the systems with higher silica content and 2 wt.% P.

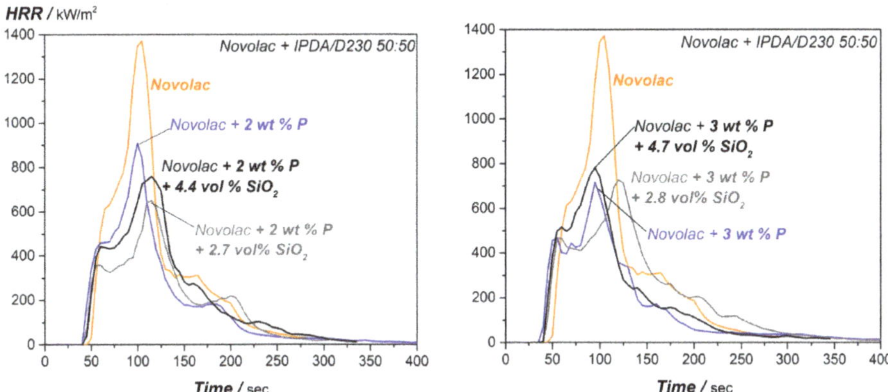

Figure 10. Cone calorimeter measurements for the modified epoxy novolac resin systems. Influence of silica content on 2 wt.% (**left**) and 3 wt.% (**right**) phosphorous modified resin.

Table 7. Time to ignition (tti) and peak heat release rate (pHRR) from cone calorimeter experiments.

	Nano-SiO$_2$ Content/vol.%	tti/sec	pHRR/kW/m^2	THR/MJ/m^2
Novolac	0.0	53.7 ± 2.9	1373.0 ± 151.6	87.7 ± 1.3
Novolac + 2 P	0.0	42.7 ± 0.6	917.7 ± 43.4	64.6 ± 1.5
Novolac + 2 P + 2.7 SiO$_2$	2.7	46.0 ± 1.0	646.6 ± 54.9	58.3 ± 1.9
Novolac + 2 P + 4.4 SiO$_2$	4.4	44.7 ± 0.6	781.0 ± 19.2	73.5 ± 2.7
Novolac + 3 P	0.0	41.0 ± 2.0	689.2 ± 43.2	57.4 ± 1.1
Novolac + 3 P + 2.8 SiO$_2$	2.8	44.7 ± 0.6	711.3 ± 30.7	76.9 ± 2.3
Novolac + 3 P + 4.7 SiO$_2$	4.7	43.3 ± 0.6	814.2 ± 28.0	63.6 ± 0.6

For deeper evaluation and understanding of the flame retarding effect of the silica and/or DOPO modified epoxy novolac systems, SEM studies of the char morphology of cone calorimeter samples are performed. The images for neat novolac, 2.3 vol.% SiO$_2$, 2 wt.% phosphorous + 2.7 vol.% SiO$_2$ and 3 wt.% phosphorous + 2.8 vol.% SiO$_2$ modified epoxy resin are represented in Figure 11 with different magnifications. The neat novolac char morphology shows rough and uneven surface (see Figure 11, upper image), and a typical carbon black like char residue structure is represented in the image with higher magnification. The morphology shows no flame retarding structure like stable char layer for instance which is also indicated in the HRR-curve in Figure 10 as the curve rises up to the maximum when the majority of the material is already decomposed and afterwards slowly declines again. DOPO modified epoxy resin shows the same char morphology as neat novolac, the flame-retardant effect is verified by the gas phase. However, the 2.3 vol.% silica modified epoxy resin shows smooth char layer with a lower number of leaks. It is estimated that silica particles induce the formation of a char layer which traps the evaporating material up to the point when the inner pressure is reaching the strength of the layer and the combustible gases escape from the generated leaks in the layer. The stable and dense SiO$_2$-layer is already proven with the Raman measurement.

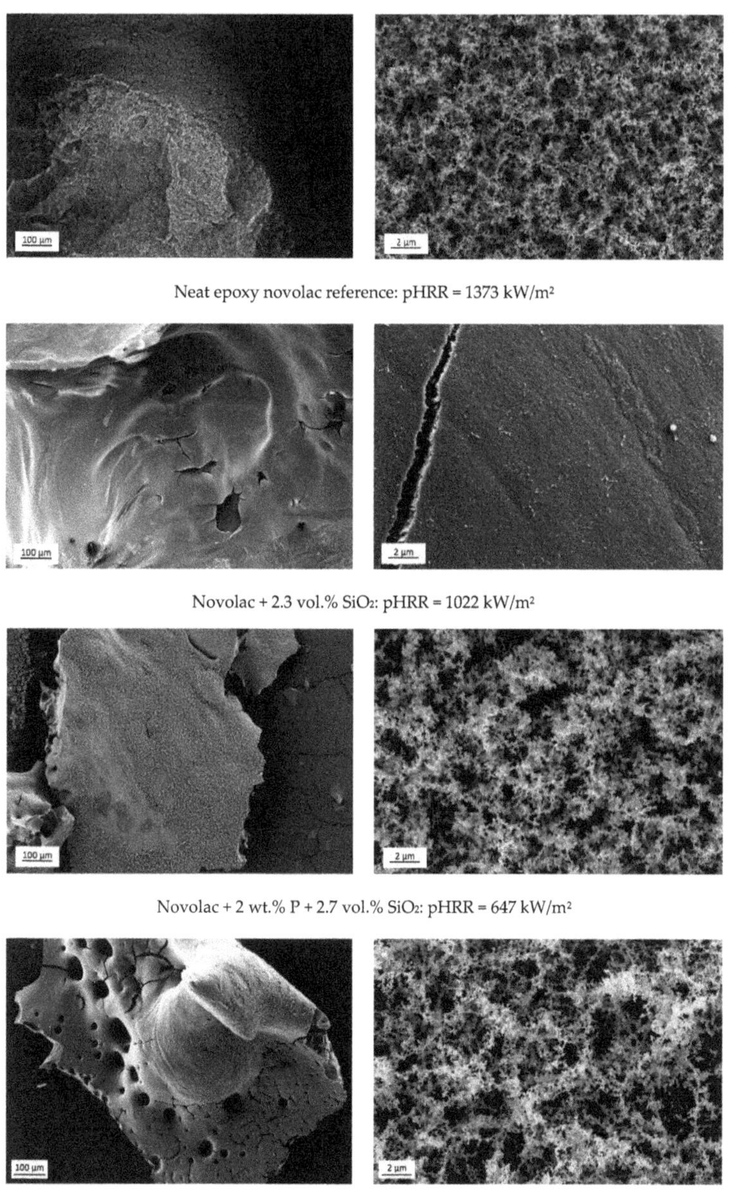

Neat epoxy novolac reference: pHRR = 1373 kW/m²

Novolac + 2.3 vol.% SiO₂: pHRR = 1022 kW/m²

Novolac + 2 wt.% P + 2.7 vol.% SiO₂: pHRR = 647 kW/m²

Novolac + 3 wt.% P + 2.8 vol.% SiO₂: p-HRR = 711 kW/m²

Figure 11. SEM images of the char morphology of cone calorimeter samples from neat novolac, 2.3 vol.% SiO_2, 2 wt.% phosphorous + 2.7 vol.% SiO_2 and 3 wt.% phosphorous + 2.8 vol.% SiO_2) modified resin, with lower (**left**) and higher magnification (**right**).

The char morphology of cone calorimeter samples with both flame-retardant additives also show the carbon black like structure at higher magnification (see Figure 11), indeed there is a dense layer visible below this morphology which is shown in the image with lower magnification. However, the 3 wt.% phosphorous and 2.8 vol.% SiO_2 modified novolac resin also shows a high number of leaks in the char layer resulting in decreased flame-retardant properties. In contrast to this, 2 wt.% phosphorous and 2.7 vol.% SiO_2 modified novolac resin, the char layer seems to be more stable as there are less leaks generated by the combusting gases shown in the SEM micrograph. The stable char layer is also indicated by the pronounced shoulder formation in the HRR-curve (see Figure 10) as it delays the release of combustible gases to the burning zone and therefore reduces the energy release of the fire. Considering the results from the TGA measurements, the resin system with 2 wt.% P + 2.7 vol.% SiO_2 shows highest deviation from the adjusted silica content (36.1%, see Table 6) which consequently demonstrates that the highest phosphorous amount is incorporated in the char layer and the thickest layer is generated. The combinational effect of the DOPO- and silica modified systems is referred to a stable and dense char layer which develops during the burning process of the material. The layer can clearly be seen in the SEM micrographs in Figure 11. Due to its high temperature stability (stable up to 1000 °C, see Figure 9), it is estimated that the residual product is highly oxidized and composed of silicone- and phosphorous-oxides, basically. Once the protective effect of the char layer reaches its limits as the gas pressure increases up to a critical value, the gas is released by the formation of slight leaks and the burning process is supported. However, it can be seen from the SEM images and the HRR-curve that protective effect of the char layer of the 2.7 vol.% nano-SiO_2 and 2 wt.% phosphorous modified resin is the most effective one.

4. Conclusions

In this study, a novel flame-retardant system containing DOPO and nano-SiO_2 modified epoxy novolac resin is evaluated and combinational effects are verified using micrograph images. The thermal, mechanical and flame-retardant properties as well as the dispersion quality, fracture and char morphology are investigated. It is shown that DOPO modified novolac resin supports agglomeration of silica particles and decreases glass transition temperature. However, the fracture toughness is enhanced with the combination of phosphorous and silica which is attributed to plastic deformation and shear yielding of the epoxy matrix which is induced by the silica particles. A higher amount of DOPO leads to the formation of agglomerates which deteriorate the shear yielding mechanisms. The decomposition behavior and flame-retardant properties are evaluated with the TGA and the cone calorimeter. The peak-HRR could be reduced from 1373.0 kW/m^2 for neat novolac up to 646.6 kW/m^2 for 2 wt.% phosphorous and 2.7 vol.% SiO_2 modified novolac resin, the time to ignition is decreased from 54 s to 46 s. It is found that there is an optimal composition for DOPO and nano-silica content regarding HRR curve which could be explained by a stable and highly oxidized char layer separating combustible gases from the burning zone. The layer is attributed to both flame retarding additives which is verified with the TGA curves as the residual mass exceeds the overall mass of the additives at higher temperatures (800–1000 °C) and results in a silicone and phosphorous-oxide containing residual product. The char structure could be verified further with SEM micrographs.

Author Contributions: Conceptualization, M.H., K.P., G.B. and R.M.; investigation, M.H., K.P., G.B.; resources and project administration, V.A. and M.M.; data curation, writing—original draft preparation and visualization, M.H.; validation, writing—review and editing, M.H., K.P., G.B. and V.A.; supervision, K.P., G.B., V.A. and M.M.

Funding: This research was funded by Forschungskuratorium Textil e.V, grant number IGF-No. 19737 N".

Acknowledgments: The research project IGF-No. 19737 N of the research associations Forschungskuratorium Textil e.V., Reinhardtstraße 12-14, 10117 Berlin, and Dechema, Theodor-Heuss-Allee 25, 60486 Frankfurt am Main, was provided via Arbeitsgemeinschaft industrieller Forschungsvereinigungen e.V. (AiF) within the PhD program of "Industrielle Gemeinschaftsforschung und –entwicklung" (IGF) of the Federal Ministry of Economics and Technology due to a resolution of the German Bundestag. The author would like to thank Josef Breu and Florian Puchtler from the department "Anorganic Chemistry I" at the University of Bayreuth for cone calorimeter measurements, Holger Schmalz (Bavarian Polymer Institute, BPI, Keylab Synthesis and Molecular Characterization) for Raman measurements and Annika Pfaffenberger from the department "Polymer Engineering" for the SEM and TEM studies. Open access charges were funded by the German Research Foundation (DFG) and the University of Bayreuth in the funding program Open Access Publishing.

References

1. Ratna, D. *Handbook of Thermoset Resins*; ISmithers: Shawbury, UK, 2009.
2. Lo, J.; Pearce, E.M. Flame-Retardant Epoxy Resins Based on Phthalide Derivative. *J. Polym. Sci.* **1984**, *22*, 1707–1715. [CrossRef]
3. Wang, C.-S.; Berman, J.R.; Walker, L.L.; Mendoza, A. Meta-Bromobiphenol Epoxy Resins: Applications in Electronic Packaging and Printed Circuit Board. *J. Appl. Polym. Sci.* **1991**, *34*, 1315–1321. [CrossRef]
4. Wang, P.; Yang, F.; Li, L.; Cai, Z. Flame-retardant properties and mechanisms of epoxy thermosets modified with two phosphorus-containing phenolic amines. *J. Appl. Polym. Sci.* **2016**, *133*, 43953–43967. [CrossRef]
5. Wang, M.; Fang, S.; Zhang, H. Study on flame retardancy of TGDDM epoxy resin blended with inherent flame-retardant epoxy ether. *High Perform. Polym.* **2017**, *30*, 318–327. [CrossRef]
6. Unlua, S.M.; Doganb, S.D.; Doganc, M. Comparative study of boron compounds and aluminum trihydroxide as flame retardant additives in epoxy resin. *Polym. Adv. Technol.* **2014**, *25*, 769–776. [CrossRef]
7. Levchik, S.V.; Weil, E.D. Review: Thermal decomposition, combustion and flame-retardancy of epoxy resins-a review of the recent literature. *Polym. Int.* **2004**, *35*, 1908–1929. [CrossRef]
8. Zhang, W.; Li, X.; Li, L.; Yang, R. Study of the synergistic effect of silicon and phosphorus on the blowing-out effect of epoxy resin composites. *Polym. Degrad. Stab.* **2012**, *97*, 1041–1048. [CrossRef]
9. Rösler, J.; Harders, H.; Bäker, M. *Mechanisches Verhalten der Werkstoffe*; Springer: Berlin, Germany, 2006.
10. Perez, R.M.; Sandler, J.K.W.; Altstädt, V.; Hoffmann, T.; Pospiech, D.; Ciesielski, M.; Döring, M. Effect of DOP-based compounds on fire retardancy, thermal stability, and mechanical properties of DGEBA cured with 4, 4'-DDS. *J. Mater. Sci.* **2006**, *41*, 341–535. [CrossRef]
11. Lin, C.H.; Wang, C.S. Novel phosphorous-containing epoxy resins Part I. Synthesis and properties. *Polymer* **2001**, *42*, 1869–1878. [CrossRef]
12. Bandi, S.; Schiraldi, D.A. Glass Transition Behavior of Clay Aerogel/Poly (vinyl alcohol) Composites. *Macromolecules* **2006**, *9*, 6537–6545. [CrossRef]
13. Borowicz, P.; Latek, M.; Rzodkiewicz, W.; Laszcz, A.; Czerwinski, A.; Ratajczak, J. Deep-ultraviolet Raman investigation of silicon oxide: Thin film on silicon substrate versus bulk material. *Adv. Nat. Sci. Nanosci. Nanotechnol.* **2012**, *3*, 2043–2050. [CrossRef]
14. Pawlyta, M.; Rouzaud, J.-N.; Duber, S. Raman microspectroscopy characterization of carbon blacks: Spectral analysis and structural information. *Carbon* **2014**, *84*, 479–490. [CrossRef]
15. Hsieh, T.H.; Kinloch, A.J.; Masania, K.; Taylor, A.C.; Sprenger, S. The mechanisms and mechanics of the toughening of epoxy polymers modified with silica nanoparticles. *Polymer* **2010**, *51*, 6284–6294. [CrossRef]
16. Johnsen, B.B.; Kinloch, A.J.; Mohammed, R.D.; Taylor, A.C.; Sprenger, S. Toughening mechanisms of nanoparticle-modified epoxy polymers. *Polymer* **2007**, *48*, 530–541. [CrossRef]
17. Liu, H.-Y.; Wang, G.; Mai, Y.-W. Cyclic fatigue crack propagation of nanoparticle modified epoxy. *Compos. Sci. Technol.* **2012**, *72*, 1530–1538. [CrossRef]
18. Dittanet, P.; Pearson, R.A. Effect of silica nanoparticle size on toughening mechanisms of filled epoxy. *Polymer* **2012**, *53*, 1890–1905. [CrossRef]
19. Kothmann, M.H.; Zeiler, R.; de Anda, A.R.; Brückner, A.; Altstädt, V. Fatigue crack propagation behaviour of epoxy resins modified with silica-nanoparticles. *Polymer* **2015**, *60*, 157–163. [CrossRef]

Article

Improving Mechanical Properties and Reaction to Fire of EVA/LLDPE Blends for Cable Applications with Melamine Triazine and Bentonite Clay

Guadalupe Sanchez-Olivares [1], Antonio Sanchez-Solis [2], Octavio Manero [2], Ricardo Pérez-Chávez [1], Mario Jaramillo [1], Jenny Alongi [3] and Federico Carosio [4,*]

[1]   CIATEC, Asociación Civil, Omega 201, Colonia Industrial Delta, León 37545, Gto., Mexico
[2]   Instituto de Investigaciones en Materiales, Universidad Nacional Autónoma de México, Avenida Universidad 3000, Ciudad de México 04510, Mexico
[3]   Dipartimento di Chimica, Università degli Studi di Milano, Via Golgi 19, 20133 Milano, Italy
[4]   Dipartimento di Scienza Applicata e Tecnologia, Politecnico di Torino-Alessandria Campus, Viale Teresa Michel 5, 15121 Alessandria, Italy
*   Correspondence: federico.carosio@polito.it; Tel.: +39-0131-229303

Received: 21 June 2019; Accepted: 24 July 2019; Published: 26 July 2019

Abstract: The high flame-retardant loading required for ethylene-vinyl acetate copolymer blends with polyethylene (EVA-PE) employed for insulation and sheathing of electric cables represents a significant limitation in processability and final mechanical properties. In this work, melamine triazine (TRZ) and modified bentonite clay have been investigated in combination with aluminum trihydroxide (ATH) for the production of EVA-PE composites with excellent fire safety and improved mechanical properties. Optimized formulations with only 120 parts per hundred resin (phr) of ATH can achieve self-extinguishing behavior according to the UL94 classification (V0 rating), as well as reduced combustion kinetics and smoke production. Mechanical property evaluation shows reduced stiffness and improved elongation at break with respect to commonly employed EVA-PE/ATH composites. The reduction in filler content also provides improved processability and cost reductions. The results presented here allow for a viable and halogen-free strategy for the preparation of high performing EVA-PE composites.

Keywords: EVA/LLDPE blend; flame retardant; wire and cable; melamine triazine; clay

---

1. Introduction

The need for safe materials represents one of the main driving forces continuously pushing the developments and advances in materials science and technology. The area of fire retardant materials is of great concern. Indeed, recent catastrophic events and present legislations clearly highlight the potential danger related to fire events, as well as the environmental and toxicological risks associated with some of the most commonly used flame-retardant chemicals. Particular fire risk is associated with electrical cables as they contain several polymeric parts (insulation, bedding, sheath) constituting fuel sources (for fire start and spread) as a consequence of arcing, excessive ohmic heating (without arcing) and external heating [1,2]. Ethylene-vinyl acetate copolymer (EVA) and EVA blends with polyethylene (EVA-PE) are among the most widely used polymers for insulation and sheathing of electric cables. Common practice is to load the polymer with high amounts (typically 60–70 wt.%) of metal hydrates such as aluminum hydroxide or magnesium hydroxide [3]. Another possible strategy to more efficiently reduce the fire and environmental risks of these materials comprises the use of nanoparticles in combination with halogen-free flame-retardant additives (FRs) [4,5]. For instance, the fire retardancy of ethylene-vinylacetate (EVA) and low-density polyethylene (LDPE) blends using organoclay in combination with either aluminum or magnesium hydroxide has been assessed by thermogravimetric analysis and cone calorimetric measurements evaluating the effect of the surface layer formed during pyrolysis of the polymer nanocomposites by numerical models [6]. The inclusion of 68 wt.% of metal hydroxides considerably reduced the heat release rate (HRR); this effect was further improved by coupling the hydroxide with 5 wt.% of organoclay. The reduction of HRR was attributed to the formation of a ceramic-like layer on the surface of unpyrolysed material in the solid phase during pyrolysis. This ceramic-like layer acts as a barrier reducing the heat and mass transfer thus resulting in a flame retardant effect. Remarkable improvements on flame retardancy by combinations of nanofiller and aluminum trihydroxide (ATH) were found in EVA [7]. These performances have been correlated to the formation of a more efficient protective barrier during combustion, thus, highlighting the beneficial role of the inclusion of relatively small amounts ($\leq$5 wt.%) of nanofiller.

Similar improvements have been observed in EVA blends for industrial cable applications by including modified bentonite clay and ATH; however, while significantly contributing to the flame retardancy properties, the presence of the nanofiller did not result in improvements in the tensile strength and elongation at break of prepared composites [8]. The detrimental effects on mechanical properties linked to the high FR loading required for the cable application of EVA and EVA blends is a well-known issue [9].

Indeed, due to the high loading required to obtain suitable FR properties for cable applications, the resulting materials are more rigid and brittle than the unmodified polymer and thus more prone to damage and cracking, eventually reducing the insulation and protection of the polymer sheath [10,11]. In order to solve this problem, research aims at the development of new and environmentally friendly flame-retardant systems capable of being efficient at loading levels below traditional amounts. Nitrogen-based compounds, such as triazines, represent a possible candidate to be incorporated in a FR system with improved efficiency [12,13]. As an example, the combination of ammonium polyphosphate and a commercial triazine derivative (poly-[2,4-(piperazine-1,4-yl)-6-(morpholine-4-yl)-1,3,5-triazine]/piperazine) has been adopted to produce a novel phosphorous nitrogen intumescent flame-retardant system for polypropylene capable of granting self-extinguishing properties at low FR content (i.e., 20 wt.%) [14]. The use of triazine derivatives as part of an intumescent system has proven to provide good effects on the fire retardancy performances of polymers [15]. However, such intumescent systems are relatively expensive and their application for electrical requirements is rather limited. In the present paper, melamine triazine has been used in combination with bentonite nanoparticles and aluminum hydroxide to produce a novel FR system for EVA-PE blends, capable of achieving excellent flame-retardant performance while preserving their mechanical properties. The developed FR systems allow for a reduction of the filler loading down to 37% with improved mechanical properties while granting FR performance suitable for electrical cables applications. This work provides a viable solution for the preparation of FR EVA-PE blends with reduced costs and improved efficiency.

2. Materials and Methods

Elvax® 460, a copolymer of ethylene vinyl acetate with 18 wt.% vinyl acetate content (2.5 g/10 min melt flow rate) from Dupont and BDL 92010 C, and linear low density polyethylene (LLDPE) (1.0 g/10 min melt flow rate) from PEMEX-Petroquímica, Coatzacoalcos, Ver., Mexico were purchased. Fusabond N493, an anhydride-modified ethylene copolymer (1.6 g/10 min melt flow), supplied by Dupont, was employed as compatibilizer. ALOLT 60DLS, aluminum trihydroxide (ATH), with average particle size d_{50} of 1.0–2.2 microns (99.5% purity and surface area of 12 m^2/g) was purchased from Mal Hungarian. MCA® PPM Triazine HF, melamine triazine (TRZ), poly-[2,4-(piperazine-1,4-yl)-6-(morpholine-4-yl)-1,3,5-triazine]/piperazin (Figure 1), was supplied by MCA Technologies GmbH. ATH and TRZ were used as halogen-free FRs. Actisil 220FF, sodium bentonite clay (55 meq/100g cationic capacity), was purchased from Clariant. Sodium bentonite clay was modified using *L*-lysine mono-chlorohydrated via ionic interchange reaction, as previously reported [16–18]. Briefly, 100 g of sodium bentonite clay was added to a water solution of 10 g *L*-lysine in 1.5 L. After 30 min stirring, the suspension was decantated and the solid was retrieved by filtration and drying of the modified bentonite clay (L-lysine loading = 3 wt.%).

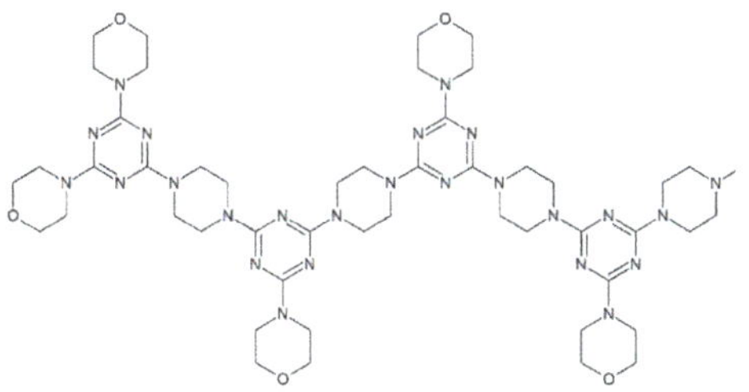

Figure 1. Chemical formula of melamine triazine.

The adopted EVA-PE blend was composed of 67 parts of EVA, 17 parts of LLDPE and 16 parts of Fusabond N493 [19–21]. This formulation constitutes the polymer matrix (EVA/LLDPE/compatibilizer) coded as E-PE. The polymer composites containing ATH, TRZ and clay were obtained in a co-rotating twin-screw extruder type SHJ 40D, with an optimized configuration for easier compounding and dispersing process as demonstrated elsewhere [22]. The extruder had a 41-mm diameter and a length/diameter ratio = 40, and 10 independent heating zones for optimal processing. A detailed description of the twin-screw configuration is reported in the Supporting Information file. Table 1 summarizes the composition of investigated composites.

Table 1. Compositions of investigated composites.

Sample	ATH [phr] *	TRZ [phr]	Clay [phr]
E-PE	-	-	-
E-PE/185ATH	185	-	-
E-PE/160ATH	160	-	-
E-PE/120ATH	120	-	-
E-PE/120ATH/20TRZ	120	20	-
E-PE/120ATH/15TRZ	120	15	-
E-PE/120ATH/10TRZ	120	10	-
E-PE/120ATH/15TRZ/1CLAY	120	15	1
E-PE/120ATH/15TRZ/3CLAY	120	15	3
E-PE/120ATH/15TRZ/5CLAY	120	15	5
E-PE/120ATH/10TRZ/1CLAY	120	10	1
E-PE/120ATH/10TRZ/3CLAY	120	10	3
E-PE/120ATH/10TRZ/5CLAY	120	10	5

* phr = parts per hundred resin.

The extrusion process was carried out at 300 rpm rotational speed with a temperature profile of 137/137/157/157/165/170/175/180/185/185 °C, from feeding zone to die (Figure S1).

Specimens for flammability, cone calorimetry and mechanical tests were produced by injection molding in a Milacron M50 machine (Milacron LLC, Karnataka, India) at 170/175/180/180 °C temperature profile, 70 mm/s injection/fill speed, 110 bar pack/hold pressure, 15 s pack/hold time and 20 s cooling time.

The morphology of E-PE blends and the corresponding composites was studied using a field-emission scanning electron microscope (SEM) JEOL JSM-7600F (JEOL, Ltd., Akishima, Japan). Specimens were prepared by using cryogenic fragile fracture and gold-coated prior to SEM analyses. Element mapping was carried out in the scanned area by energy dispersive spectroscopy (EDS, Oxford Instruments, Concord, MA, USA). Thermal stability was evaluated by thermogravimetric analysis by a TA-Instrument Q-550 equipment (TA-Instrument, Inc., New Castle, DE, USA) at a heating rate of 10 °C/min from 25 °C up to 800 °C, under argon and air atmospheres. The sample size was 10 ±1 mg, the experimental error was ±1 °C and ±0.01 wt.%. Flammability was assessed following UL94 vertical classification tests according to the ASTM D3801-19 standard [23], using specimens with dimensions of $127 \times 12.7 \times 3.1$ mm^3. Combustion behavior under forced combustion was investigated by cone calorimetry (Fire Testing Technology). Specimens ($100 \times 100 \times 3$ mm^3) were exposed to a 35 kW/m^2 radiative heat flux in horizontal configuration. Average values concerning time to ignition (TTI), peak of heat release rate (pkHRR), total heat release (THR), maximum average rate of heat emission (MARHE), total smoke release (TSR) and final residue were evaluated and are presented with their experimental deviations. Measurements were performed four times for each formulation. Prior to flammability and forced combustion tests, all specimens were conditioned in a climatic chamber (23 ±1 °C 50% relative humidity) for 48 h. Tensile tests were carried out using an Instron Universal testing machine 5565 model (Instron Corp., Norwood, MA, USA) at a crosshead speed of 50 mm/min at 25 ±2 °C and type I specimen dimensions, following the ASTM D638 standard [24]. At least five specimens were tested for each sample, and the average value for Young modulus, tensile strength, elongation at break and tenacity is reported.

The rheological behavior of the composites was measured in a strain-controlled Ares G2 TA-Instrument (TA-Instrument, Inc., New Castle, DE, USA) rheometer using parallel plates of 25 mm diameter. All tests were performed at 195 °C under small amplitude oscillatory shear flow (SAOS). The dynamic frequency sweep mode was carried out in linear viscoelastic regimen with a strain of 1% from 0.1 to 100 rad/s.

3. Results and Discussion

3.1. Morphology

The morphology of the prepared composites was studied by scanning electron microscopy (SEM). Figure 2 displays SEM micrographs of the neat polymer matrix (E-PE blend, Figure 2A) and selected corresponding composites containing ATH, ATH/TRZ and ATH/TRZ/clay (Figure 2B,D).

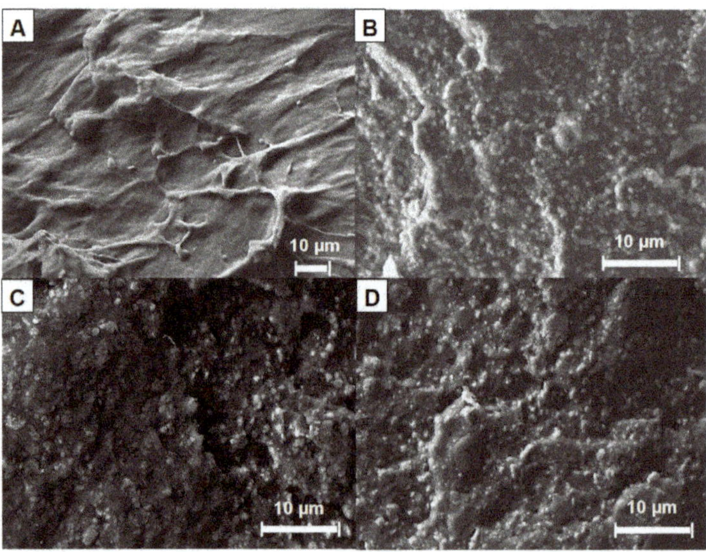

Figure 2. Scanning electron microscopy (SEM) micrographs of fractured surface for: (**A**) E-PE, (**B**) E-PE/185ATH, (**C**) E-PE/120ATH/15TRZ, and (**D**) E-PE/120ATH/15TRZ/3clay.

E-PE polymer blend shows a ductile fracture characterized by a deformed surface and a continuous pattern; no phase separation or droplets are observed for this blend (Figure 2A). This is ascribed to the use of the compatibilizer (anhydride-modified ethylene copolymer) as it is well-known that blends based on ethylene-vinyl acetate copolymer and polyethylene (i.e., LDPE or LLDPE) are immiscible [25]. The absence of a compatibilizer would result in phase separation and polymer droplet formation as a function of the interfacial tension and viscosity ratio between the EVA and PE [26]. The inclusion of ATH deeply modifies the resulting morphology in a fragile fracture attributed to the presence of the filler (Figure 2B). ATH particles, with dimensions ranging from sub micronic up to 4 μm, show a good distribution and dispersion within the polymer matrix even at such high filler loading (185 phr, 65 wt.%). This is ascribed to the screw configuration and high rotational speed employed during the melt extrusion process. A similar morphology is observed for composites containing TRZ and TRZ/clay with reduced ATH content (Figure 2C,D), thus highlighting no substantial changes in the distribution of ATH in the presence of the other additives. The distribution of clay within the composites was further evaluated by elemental mapping; Figure 3 reports the aluminum and silicon elemental analysis of E-PE/120ATH/10TRZ/5clay.

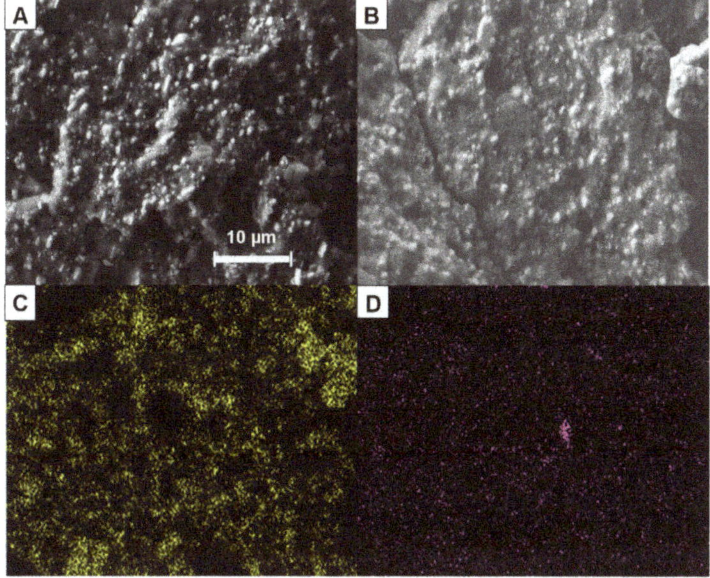

Figure 3. (**A**) SEM micrograph of fractured surface of E-PE/120ATH/10TRZ/5clay composite, (**B**) scanned area of E-PE/120ATH/10TRZ/5clay composite for elemental analysis, (**C**) aluminum and (**D**) Si mapping.

According to the silicon elemental analysis, clay particles exhibit a homogenous dispersion and distribution. No agglomerates are observed, likely due to the low clay content within the composites (i.e., 5 phr, 2 wt.%). Furthermore, Al distribution is similar to that of E-PE/120ATH (Figure S2) highlighting that the presence of TRZ and clay does not alter ATH distribution and dispersion.

3.2. Rheological Properties

Rheological properties of the E-PE blend and composites included continuous simple and small amplitude oscillatory shear flow. This test provides information on the dispersion of the filler within the polymer matrix. Figure 4 depicts shear viscosity as a function of shear rate for E-PE blend and investigated composites.

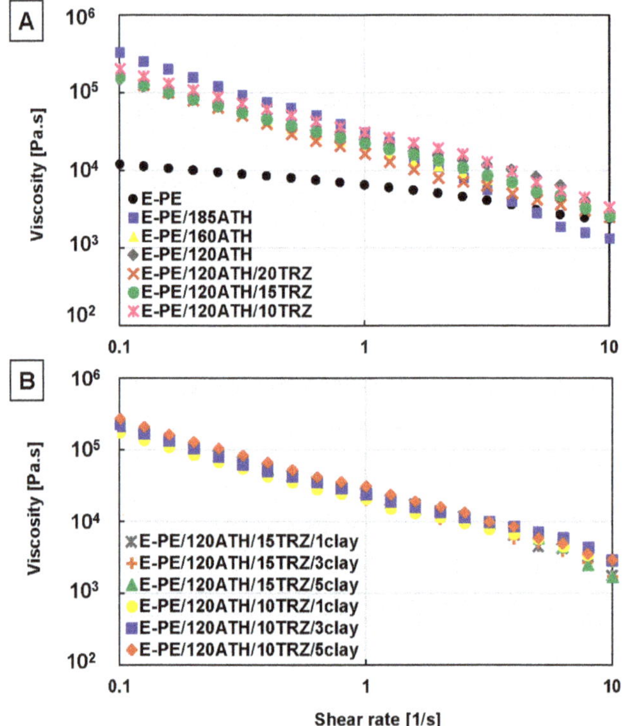

Figure 4. Simple shear viscosity as a function of shear rate for: (**A**) E-PE blend, E-PE/ATH and E-PE/120ATH/TRZ composites using ATH and TRZ at different content, and (**B**) E-PE/120ATH/15TRZ/clay and E-PE/120ATH/10TRZ/clay composites varying clay content.

In Figure 4A, the shear viscosity of the polymer matrix (E-PE blend) presents a nearly Newtonian-like behavior at low shear rates (0.1–1.0 1/s). Newtonian-like behavior corresponds to the region in the flow curve where the viscosity becomes independent of the shear rate (i.e., a constant viscosity). On the other hand, E-PE blend presents moderate shear thinning behavior at high shear rate (1.0–10.0 1/s). By including ATH and TRZ in the formulation, the viscosity increases at low shear rates (0.1–1.0 1/s), reaching the maximum value for E-PE/185ATH. Nevertheless, at high shear rates (1.0–10 1/s) the viscosity of E-PE/185ATH composite decreases similarly to the rest of other formulations. All composites exhibit the typical feature of shear-thinning behavior with no plateau region observed over the studied shear rate range. Such behavior has been ascribed to a good dispersion of both flame retardants (ATH and TRZ) within the polymer matrix (Figure 4A) [27]. Composites containing clay show a remarkable shear-thinning behavior in the whole shear rate range (Figure 4B). Such behavior implies low shear viscosity at high shear rate, an important characteristic for the easy processing of these composites. In addition, the storage modulus was also evaluated, and its plot as function of angular frequency for E-PE blend and prepared composites is reported in Figure 5.

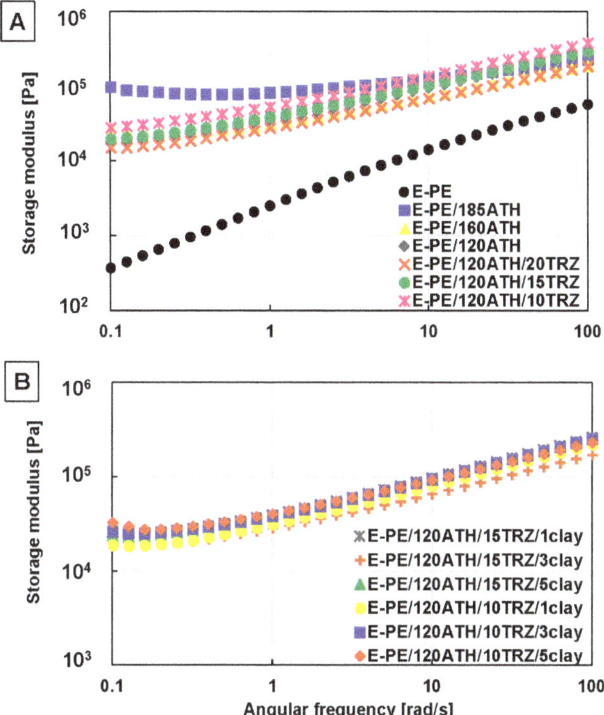

Figure 5. Storage modulus as function of angular frequency in small amplitude oscillatory shear flow (SAOS) flow test for: (**A**) E-PE blend, E-PE/ATH and E-PE/120ATH/TRZ composites using ATH and TRZ at different content, and (**B**) E-PE/120ATH/15TRZ/clay and E-PE/120ATH/10TRZ/clay composites varying clay content.

The neat E-PE blend displays a storage modulus with a constant slope over the entire frequency range, indicating non-terminal flow behavior characteristic of pure polymers [28]. Composites containing ATH and ATH/TRZ show a pronounced solid-like behavior at low frequency (0.1–1.0 rad/s). In particular, the storage modulus of E-PE/185ATH exhibited the highest value likely due to the high particle loading (Figure 5A). On the other hand, the presence of TRZ does not influence the storage modulus, as all TRZ containing composites disclose a similar slope in the whole frequency range with respect to E-PE/120ATH. This result indicates the absence of interactions between the filler particles in the flow stage. A similar behavior is observed for clay containing formulations (Figure 5B). The performed rheological measurements suggest a good dispersion of flame retardant additives within the polymer matrix, further confirming previous SEM observations.

3.3. Thermal Stability

The thermal stability of neat components and prepared composites under inert and oxidative atmosphere has been evaluated by thermogravimetric analyses in argon and air, respectively. As far as neat components are concerned (Figure S3 and Table S1), neat ATH yields a weight loss ascribed to its dehydration with consequent water release within a 200–300 °C range. Bentonite clay shows a slow and constant weight loss associated to its dehydration from the interlayer space and cavities, and cation hydration spheres, as well as dihydroxylation at high temperatures [29,30]. On the other hand, TRZ shows a more complicated degradation path associated to melamine gradual condensation releasing melam, melem and melon products due to ammonia elimination [31].

Figure 6 displays TG and dTG curves, and Table 2 discloses the collected data of the most representative samples. The complete curves and thermal data are reported in Figures S4–S6 and Tables S2–S4, respectively.

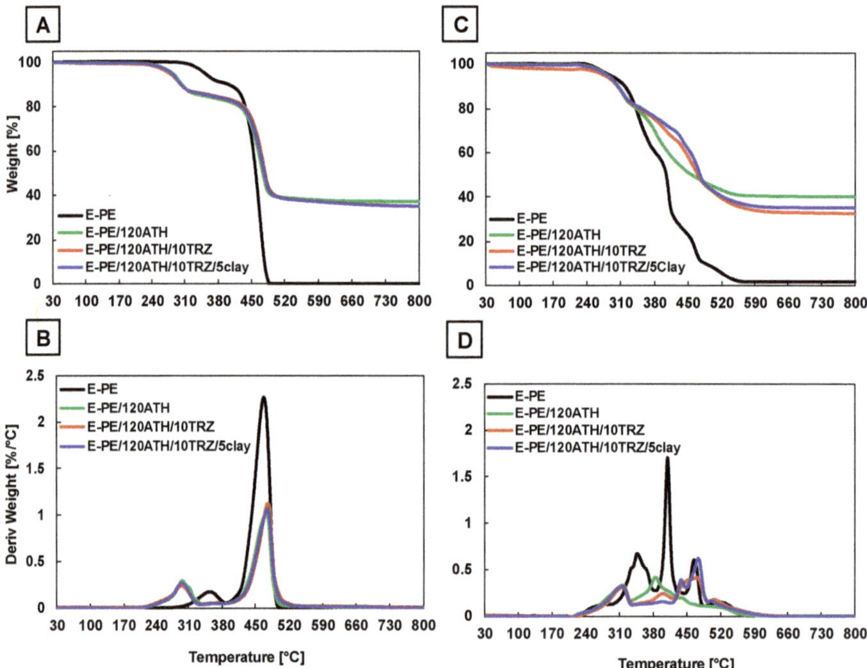

Figure 6. TG and dTG curves of E-PE, E-PE/120ATH, E-PE/120ATH/10TRZ and E-PE/120ATH/ 10TRZ/5clay composites. (**A,B**) curves in argon, and (**C,D**) curves in air atmospheres.

Table 2. Thermal data of polymer E-PE, E-PE/120ATH, E-PE/120ATH/10TRZ and E-PE/120ATH/ 10TRZ/5clay composites by thermogravimetric analyses.

Sample	Argon			Air				
	* T_{max} [°C]	Deriv. Mass [%/°C]	Residue at 800 °C [%]	* $T_{max\,1}$ [°C]	Deriv. Mass $_1$ [%/°C]	* $T_{max\,2}$ [°C]	Deriv. Mass $_2$ [%/°C]	Residue at 800 °C [%]
E-PE	467	2.27	0.0	346	0.67	410	1.69	1.5
E-PE/120ATH	472	1.02	37.2	320	0.35	385	0.48	40.2
E-PE/120ATH/10TRZ	476	1.13	35.2	318	0.31	470	0.42	32.4
E-PE/120ATH/10TRZ/5clay	476	1.06	34.8	315	0.33	467	0.49	34.9

* From derivative curves.

Under non-oxidative conditions, the E-PE blend decomposes in two steps (Figure 6A,B). The first occurs at nearly to 350 °C and corresponds to the de-acylation of the vinyl acetate groups in EVA [31,32]. The second step, associated with the highest weight loss, takes place between 400 and 500 °C as a result of EVA unsaturated backbone and PE hydrocarbon chains decomposition, leaving no residue at 800 °C [33,34]. The presence of ATH is responsible for an anticipated degradation due to water release, and the formation of an inorganic barrier that partially slows down the E-PE decomposition, as observed in the dTG curves (Figure 6 and Figure S4). The final residue consists of aluminum oxide, and increases as the ATH loading increases. TRZ and clay do not substantially modify this behavior likely due to the lower content with respect to ATH, as confirmed in samples containing different amounts of TRZ and clay (Figures S5 and S6).

Under thermo-oxidative conditions, EVA de-acylation still occurs within the first decomposition step. Subsequently, the presence oxygen results in two separate weight loss steps in the 350–470 °C range related to the different thermo-oxidation of EVA and PE that produces a 10 wt.% residue, eventually oxidized above 500 °C (Figure 6C,D) [35]. The barrier produced by ATH limits oxygen diffusion and results in delayed and reduced degradation kinetics above 350 °C. This effect is clearly visible in the TG and dTG curves of Figure 6C,D. The presence of TRZ and clay improve the efficiency of the produced barrier as observable in the residues at 450 °C (i.e., 51%, 59% and 62%, for E-PE/120ATH, E-PE/120ATH/10TRZ and E-PE/120ATH/10TRZ/5Clay, respectively). TRZ is effective only at 15 and 20 phr, whereas different clay content does not change this effect, as reported in Figures S5 and S6.

3.4. Flammability

The flammability of E-PE blend and the investigated composites using ATH, TRZ additives and modified bentonite was assessed by UL94 vertical classification. This test evaluates the reaction of prepared materials when subjected to a direct flame application, thus providing information on their ability to start a fire. Table 3 reports UL94-V classification of the investigated materials and Figure 7 collects digital images of some specimens at the end of the test.

Table 3. Flammability results of E-PE blend and the investigated composites following the UL94 vertical configuration method.

Sample	$t\,1 \pm \sigma$	$t\,2 \pm \sigma$	UL94 Classification	Burning Characteristics
E-PE	>60	-	n.c. *	Intense melt dripping
E-PE/185ATH	-	3 ± 1	V0	No melt dripping
E-PE/160ATH	-	85 ± 26	n.c.	Moderated melt dripping
E-PE/120ATH	-	57 ± 35	n.c.	Flaming droplets
E-PE/120ATH/20TRZ	-	3 ± 1	V0	No melt dripping
E-PE/120ATH/15TRZ	41 ± 78	105 ± 30	n.c.	Intense melt dripping
E-PE/120ATH/10TRZ	7 ± 8	77 ± 40	n.c.	Intense melt dripping
E-PE/120ATH/15TRZ/1CLAY	-	7 ± 5	V1	No melt dripping
E-PE/120ATH/15TRZ/3CLAY	-	4 ± 2	V0	No melt dripping
E-PE/120ATH/15TRZ/5CLAY	-	4 ± 2	V0	No melt dripping
E-PE/120ATH/10TRZ/1CLAY	-	83 ± 39	n.c.	Intense melt dripping
E-PE/120ATH/10TRZ/3CLAY	-	9 ± 3	V1	No melt dripping
E-PE/120ATH/10TRZ/5CLAY	-	4 ± 3	V0	No melt dripping

* n.c.: not classifiable. Note: the occurrence of intense melt dripping in n.c. samples are responsible for a large standard deviation as this might cause the specimen to self-extinguish at random times.

As is well-known, the E-PE blend is a highly flammable material. Indeed, upon flame application the sample starts to burn vigorously with the formation of flaming droplets that may extinguish the flame prior to the complete combustion of the sample, as reported in Figure 7A. This is highly undesirable as this phenomenon can easily spread the fire to other ignitable materials in a real fire scenario, thus resulting in a serious fire threat. The inclusion of ATH considerably changes the burning behavior of the composite (Figure 7B,C). Indeed, at 185 phr no ignition is observed after the first flame application while the second flame application results in short burning times (<5 s) granting the maximum rating for this test: V0 classification (Figure 7A). This is due to the formation of a protective inorganic barrier as ATH accumulates on the surface of the specimen exposed to the flame; the release of water also provides beneficial effects lowering the temperature of the flame and diluting volatiles. Reducing the content of ATH to 160 and 120 phr (Figure 7C) compromises the performances of the composites with a downgrade to not classifiable rating due to the presence of melt dripping. These results clearly confirm the mandatory need for very high ATH content in order to achieve good flame retardant effects, as already reported in the literature. To overcome this problem, TRZ and clay have been added to the formulation containing 120 phr of ATH (Figure 7D,E). The inclusion of 20 phr of TRZ alone allows for the maximum rating pairing the results of the E-PE/185ATH composites (see Figure 7D and Table 3). A reduction to 15 and 10 phr does not grant similar performances and results in extensive melt dripping (Figure 7E,G) and prolonged burning times (>60 s). Such results are improved by the addition of clay at either 3 or 5 phr, achieving the highest rating with TRZ at both 15 and 10 phr (Figure 7F,H). It is worth highlighting the beneficial role of modified bentonite that is capable of considerable improvements in the flame-retardant performances at relatively low loadings (i.e., E-PE/120ATH/10TRZ/5clay). Such results can be related to the good distribution and dispersion of the clay during processing. This is deemed to have a fundamental role in the achieved flame retardancy properties, as it allows the clay to substantially improve the efficiency of the barrier produced by ATH [7,36]. From the above results, the TRZ/clay combination allows for substantial reductions in the total filler loadings while maintaining fire safety.

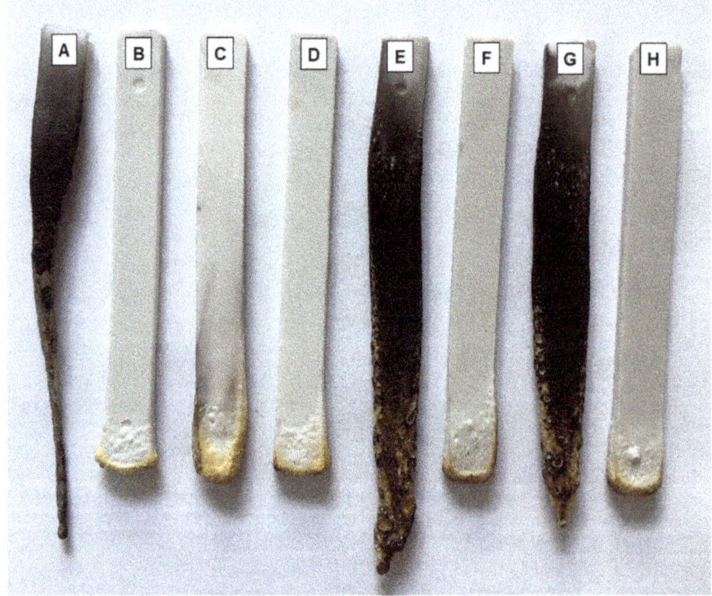

Figure 7. Digital pictures of specimens after UL94-V tests: (**A**) E-PE, (**B**) E-PE/185ATH, (**C**) E-PE/120ATH, (**D**) E-PE/120ATH/20TRZ, (**E**) E-PE/120ATH/15TRZ, (**F**) E-PE/120ATH/15TRZ/3clay, (**G**) E-PE/120ATH/10TRZ, and (**H**) E-PE/120ATH/10TRZ/5clay composites.

3.5. Burning Behavior under Forced Flaming Combustion

Cone calorimetry was employed to evaluate the reaction of prepared composites to the exposure to a heat flux typical of developing fires (i.e., 35 kW/m^2). For this test, samples have been selected on the basis of flammability results and total filler loading in order to test the more efficient formulations along with their reference material. During the test, as a consequence of the heat flux exposure, the sample starts degrading and releasing flammable volatiles that are ignited by a spark positioned above the samples. Once ignition occurs, the instrument evaluates all parameters linked to heat and smoke release. The main parameter is the heat release rate, which as function of time, is reported in Figure 8. Table 4 collects the complete set of parameters for each composite.

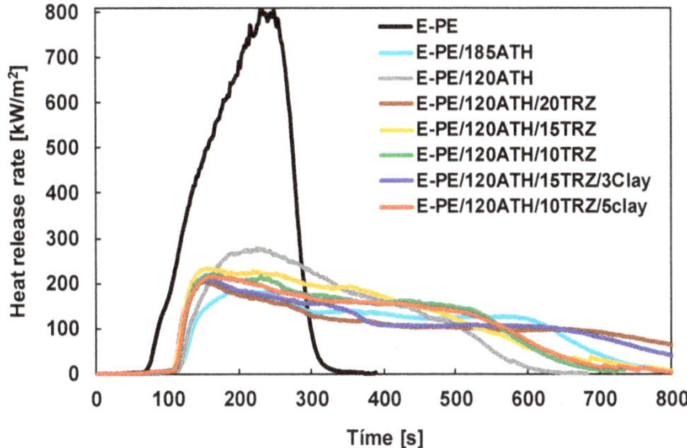

Figure 8. Heat release rate curves as a function of time for: E-PE, E-PE/185ATH, E-PE/120ATH, E-PE/120ATH/20TRZ, E-PE/120ATH/15TRZ, E-PE/120ATH/15TRZ/3clay, E-PE/120ATH/10TRZ, and E-PE/120ATH/10TRZ/5clay composites.

Table 4. Combustion data results of E-PE and some corresponding composites by cone calorimetry.

Sample	TTI [s]	pkHRR [kW/m²]	THR [MJ/m²]	MARHE [kW/m²]	TSR [m²/m²]	Residue [%]
E-PE	62 ± 4	850 ± 59	110 ± 1	379 ± 15	1178 ± 33	0
E-PE/185ATH	111 ± 5	186 ± 2	77 ± 4	112 ± 2	570 ± 53	44 ± 1
E-PE/120ATH	107 ± 2	281 ± 14	80 ± 9	155 ± 5	907 ± 62	37 ± 1
E-PE/120ATH/20TRZ	98 ± 3	210 ± 5	88 ± 1	109 ± 4	725 ± 44	35 ± 1
E-PE/120ATH/15TRZ	99 ± 5	237 ± 13	86 ± 5	147 ± 9	951 ± 68	33 ± 1
E-PE/120ATH/10TRZ	103 ± 4	223 ± 15	83 ± 5	138 ± 10	945 ± 42	35 ± 1
E-PE/120ATH/15TRZ/3clay	101 ± 4	212 ± 12	86 ± 3	116 ± 3	843 ± 47	35 ± 1
E-PE/120ATH/10TRZ/5clay	101 ± 3	218 ± 9	82 ± 6	133 ± 2	806 ± 43	36 ± 1

An apparent flame-retardant effect can be achieved by including 185 phr of ATH with considerable reduction in heat release values (pkHRR, THR and MARHE reduced by 78%, 30% and 70%, respectively) as well as smoke production (TSR reduced by 52%). This result is related to both the reduced amount of polymer matrix in the composite and to the barrier and water release effect produced by ATH [37]. The produced barrier, clearly visible from the digital pictures of the residues reported in Figure S7, hinders volatile release and limits heat transmission and mass transfer from the flame to the polymer, resulting in reductions of combustion kinetics as well as smoke production. The released water can dilute smoke by reducing its optical density while simultaneously lowering the flame temperature. Reducing the content of the hydroxide to 120 phr maintains good flame retardant properties with the most apparent detrimental effect on TSR values likely due to the production of an inefficient barrier during combustion and to the lesser release of water. On the other hand, it should be pointed out that by reducing the ATH content, the amount of combustible polymer is inevitably increased, thus providing an additional challenge for the developed formulations.

As observed from flammability results, the inclusion of TRZ helps in improving the properties of the 120 phr ATH formulation from the assessment of pkHRR and TSR reductions. This can be ascribed to the mode of action of TRZ that, as for other melamine derivatives, shows mostly diluting and cooling effects in the gas phase [38]. The beneficial effect of TRZ is remarkable only at 20 phr while lower loadings only partially improve the pkHRR reduction (see Table 4). Further improvements of the formulations containing 15 and 10 phr of TRZ can be achieved by incorporating bentonite clay. The ability of clay in promoting the formation of a more efficient barrier to volatiles and heat transfer allows for further reducing pkHRR and TSR values [39,40], as reported in Table 4 mostly matching the results of E-PE/120ATH/20TRZ. The evaluation, by optical microscopy, of the top surface of the residues collected at the end of the test (Figure S8) show no apparent differences between each formulation. A compact and brittle inorganic layer mostly resulting from the cumulation of aluminum oxide at the polymer/flame interface is observed. The internal structure of the residue has also been investigated. To this aim, small pieces have been collected from the main structure and tilted in order to make the internal structure visible (Figure S9). Differences in macroscopic morphology can be easily detected. Indeed, while formulations containing 185 and 120 phr of ATH yielded a quite dense structure, the presence of TRZ produced porous structures with pore number and distribution proportional to TRZ content. This can be ascribed to the release of volatiles by TRZ and helps in improving the heat shielding properties of the produced protective layer that benefits from the reinforcing effect of clay [31,36]. From an overall point of view, the addition of TRZ and clay compensates for the reduced ATH content, thus providing a valuable strategy to simultaneously reduce the FR loading while guaranteeing considerable FR performances. Indeed, the observed reductions in combustion parameters (pkHRR, THR and MARHE reduced by 74%, 25% and 65%, respectively) and smoke production (TSR reduced by 32%), ensures the fire safety of the ATH/TRZ/clay formulations.

3.6. Mechanical Properties

The impact of the flame-retardant formulation on the mechanical properties of the E-PE blend was assessed by tensile tests [24]. Table 5 collects Young's modulus, tensile strength, elongation at break and tenacity of prepared composites.

Table 5. Mechanical properties of E-PE blend and corresponding composites.

Sample	Young's Modulus [MPa]	Tensile Strength [MPa]	Elongation at Break [%]	Tenacity [MPa]
E-PE	27 ± 1	7.0 ± 0.3	478 ± 26	27 ± 2
E-PE/185ATH	92 ± 5	13.0 ± 0.4	101 ± 8	11 ± 1
E-PE/160ATH	68 ± 1	12.0 ± 0.2	147 ± 9	14 ± 1
E-PE/120ATH	58 ± 1	10.0 ± 0.2	165 ± 11	13 ± 1
E-PE/120ATH/20TRZ	58 ± 2	7.0 ± 0.4	115 ± 6	6 ± 1
E-PE/120ATH/15TRZ	61 ± 2	11.0 ± 0.1	180 ± 7	16 ± 1
E-PE/120ATH/10TRZ	68 ± 2	12.0 ± 0.4	167 ± 9	16 ± 1
E-PE/120ATH/15TRZ/1CLAY	69 ± 1	11.0 ± 0.2f	146 ± 11	13 ± 1
E-PE/120ATH/15TRZ/3CLAY	74 ± 3	11.0 ± 0.2	137 ± 4	12 ± 0
E-PE/120ATH/15TRZ/5CLAY	77 ± 2	11.0 ± 0.3	113 ± 5	10 ± 1
E-PE/120ATH/10TRZ/1CLAY	74 ± 2	12.0 ± 0.4	130 ± 9	13 ± 1
E-PE/120ATH/10TRZ/3CLAY	70 ± 2	12.0 ± 0.3	142 ± 6	13 ± 1
E-PE/120ATH/10TRZ/5CLAY	70 ± 1	10.0 ± 0.2	137 ± 8	11 ± 1

As is well-known, the high contents of FR additives needed to ensure safety inevitably result in substantial changes of the mechanical properties of the polymer matrix, increasing modulus and tensile strength while reducing elongation at break and tenacity [9,41]. Such behavior is observed for formulation containing 185 phr of ATH that show increased stiffness (Young's modulus and tensile strength up to 92 and 13 MPa, respectively) and reduced deformability (elongation at break reduced from 478% to 101%). The reduction of ATH content from 185 to 120 phr partially limits this phenomenon (elongation at break is improved), but has the unwanted result of considerably limiting the fire safety of the prepared materials as demonstrated by flammability and cone testing. The inclusion of TRZ at 20 phr further increases the stiffness of the materials, while reducing its content to 15 and 10 phr partially improves the elongation at break and tenacity of the formulations with respect to E-PE/120ATH/20TRZ. Similarly, bentonite clay does not improve deformability with respect to the formulations containing ATH and TRZ. However, it should be pointed out that the elongation at break displayed by formulations containing TRZ and clay is always superior to E-PE/185ATH (i.e., 137% vs. 101%), which is the reference material as far as fire protection is concerned. Such improvements can be mainly ascribed to the reduced additive content. This indicates that composites such as E-PE/120ATH/15TRZ/3clay and E-PE/120ATH/10TRZ/5clay can improve on the overall deformability of the materials while still maintaining the required flame retardant properties. In order to evaluate the potential economic impact of the performed formulations, a simple cost-benefit analysis has been performed evaluating the cost of raw materials employed in the most performing formulations (Table 6).

Table 6. Cost of the total loading for composite classified as V0 following the UL94 vertical configuration.

Sample	Formulation Composition	* Cost of Formulation/kg of E-PE [USD/kg]	* Cost of Formulation/m^3 of E-PE [USD/m^3]
E-PE/185ATH	185 phr ATH	7.3	6.8
E-PE/120ATH/20TRZ	120 phr ATH+20 phr TRZ	7.7	7.1
E-PE/120ATH/15TRZ/3clay	120 phr ATH+15 phr TRZ+3 clay	7.1	6.6
E-PE/120ATH/10TRZ/5clay	120 phr ATH+10 phr TRZ+5 clay	6.4	5.9

* Cost estimated for ATH: 3.95 USD/kg, TRZ: 15.00 USD/kg, clay: 5.00 USD/kg.

It is apparent that the optimized E-PE/120ATH/10TRZ/5clay helps in saving up to 13% of the costs, with respect to the E-PE/185ATH reference sample (both to the same V0 classification according to the UL94 vertical configuration), thus making this formulation the most appealing from an industrial point of view.

4. Conclusions

In this work, melamine triazine and bentonite clay have been employed as novel flame-retardant additives for ethylene-vinyl acetate copolymer blends with polyethylene loaded with reduced amounts (i.e., 120 phr) of conventional aluminum trihydroxide particles. The aim was to maintain excellent flame-retardant properties, comparable with those of conventionally employed E-PE composites at high filler loading (i.e., 185 phr), while preserving mechanical properties. Different contents of TRZ and clay at fixed ATH content were prepared and thoroughly investigated from the morphology, rheology, thermal stability, flame retardancy and mechanical properties point of view. Optimized E-PE formulations grant self-extinguishing behavior during flammability tests in the vertical configuration, reaching the highest classification rating (V0) while E-PE/120ATH composites fail the test (not classifiable). The presence of TRZ and clay improves the efficiency of the protective barrier produced by ATH during combustion. This was also confirmed by cone calorimetry where samples containing TRZ and clay were capable of further reducing combustion kinetics (−23% in pkHRR) and smoke production (−11% in TSR) with respect to the E-PE/120ATH reference. Mechanical properties showed significant improvements as compared with conventional formulations (i.e., E-PE/185ATH) with reduced stiffness and improved elongation at break.

The ability of preserving mechanical properties while still achieving high flame-retardant performances in combination with easier processing conditions and reduced costs make the composites developed in this work highly promising and attractive solutions for further industrial exploitation.

Supplementary Materials: The following are available online at http://www.mdpi.com/1996-1944/12/15/2393/s1, Figure S1: Digital pictures of the screw configuration. 1-5 kneading element blocks, Figure S2: (A) Area of E-PE/120ATH composite analyzed by SEM for elemental analysis. (B) Aluminum mapping, Figure S3: TG and dTG curves of ATH, TRZ, and clay. (A,B) curves in argon, and (C,D) curves in air atmospheres, Figure S4: TG and dTG curves of E-PE/ATH composites varying ATH content. (A,B) curves in argon, and (C,D) curves in air atmospheres, Figure S5: TG and dTG curves of E-PE/120ATH/TRZ composites varying TRZ amount. (A,B) curves in argon, and (C,D) curves in air atmospheres, Figure S6: TG and dTG curves of E-PE/120ATH/15TRZ/clay and E-PE/120ATH/10TRZ/clay composites varying the content of modified bentonite. (A,B) curves in argon, and (C,D) curves in air atmospheres, Figure S7: Digital pictures of residues after cone calorimetry tests for: (A) E-PE, (B) E-PE/185ATH, (C) E-PE/120ATH, (D) E-PE/120ATH/20TRZ, (E) E-PE/120ATH/15TRZ, (F) E-PE/120ATH/15TRZ/3clay, (G) E-PE/120ATH/10TRZ, and (H) E-PE/120ATH/10TRZ/5clay composites. Figure S8: Optical microscopy pictures of the surface of the residues after cone calorimetry tests for: (A) E-PE/185ATH, (B) E-PE/120ATH, (C) E-PE/120ATH/20TRZ, (D) E-PE/120ATH/15TRZ, (E) E-PE/120ATH/10TRZ, (F) E-PE/120ATH/15TRZ/3clay, (G) E-PE/120ATH/10TRZ/5clay composites, Figure S9: Digital pictures of the internal portion of the residues after cone calorimetry tests for: (A) E-PE/185ATH, (B) E-PE/120ATH, (C) E-PE/120ATH/20TRZ, (D) E-PE/120ATH/15TRZ, (E) E-PE/120ATH/10TRZ, (F) E-PE/120ATH/15TRZ/3clay, (G) E-PE/120ATH/10TRZ/5clay composites. Table S1: Thermal data of ATH, melamine triazine (TRZ) and modified bentonite (clay) in argon and air atmospheres, Table S2: Thermal data of E-PE/ATH composites using different content of ATH by thermogravimetric analysis, Table S3: Thermal data of E-PE/120ATH/TRZ composites varying TRZ content by thermogravimetric analysis, Table S4: Thermal data of E-PE/120ATH/15TRZ/clay and E-PE/120ATH/10TRZ/clay composites varying bentonite content by thermogravimetric analysis.

Author Contributions: Conceptualization, G.S.-O., A.S.-S. and O.M.; methodology, G.S.-O. and A.S.-S.; investigation, R.P.-C. and M.J.; validation, G.S.-O. and A.S.-S; visualization, G.S.-O. and F.C.; writing and original draft preparation, G.S.-O., J.A. and F.C.; writing, review and editing, F.C., J.A. and O.M.; funding acquisition, G.S.-O. and O.M.

Funding: This work was funded by the National Council of Science and Technology of Mexico (CONACYT/235880 project).

Acknowledgments: The authors thank G. Beyer for the recommendation of the materials used in this work. We would like to acknowledge P. Rodríguez Garcia for thermogravimetric measurements, O. Novelo Peralta for SEM observations, and F. Cuttica for cone calorimetry tests.

Conflicts of Interest: The authors declare no conflicts of interest.

References

1. Meinier, R.; Sonnier, R. Fire behavior of halogen-free flame retardant electrical cables with the cone calorimeter. *J. Hazard. Mater.* **2018**, *135*, 3069–3083. [CrossRef] [PubMed]

2. Martinka, J.; Rantuch, P.; Sulová, J.; Martinka, F. Assessing the fire risk of electrical cables using a cone calorimeter. *J. Therm. Anal. Calorim.* **2019**, *342*, 306–316. [CrossRef]

3. Girardin, B.; Fontaine, G.; Duquesne, S.; Försth, M.; Bourbigot, S. Characterization of Thermo-Physical Properties of EVA/ATH: Application to Gasification Experiments and Pyrolysis Modeling. *Materials* **2015**, *8*, 7837–7863. [CrossRef] [PubMed]

4. Morgan, A.B.; Wilkie, C.A. Practical issues and future trends in polymer nanocomposite flammability research. In *Flame Retardant Polymer Nanocomposites*; Morgan, A.B., Wilkie, C.A., Eds.; John Wiley & Sons, Inc.: Hoboken, NJ, USA, 2007; pp. 355–356.

5. Kiliaris, P.; Papaspyrides, C.D. Polymer/layered silicate (clay) nanocomposites: An overview of flame retardancy. *Prog. Polym. Sci.* **2010**, *35*, 902–958. [CrossRef]

6. Zhang, J.; Hereid, J.; Hagen, M.; Bakirtzis, D.; Delichatsios, M.A.; Fina, A.; Castrovinci, A.; Camino, G.; Samyn, F.; Bourbigot, S. Effects of nanoclay and fire retardants on fire retardancy of a polymer blend of EVAandLDPE. *Fire Saf. J.* **2009**, *44*, 504–513. [CrossRef]

7. Beyer, G. Flame Retardant Properties of EVA-nanocomposites and Improvements by Combination of Nanofillers with Aluminium Trihydrate. *Fire Mater.* **2001**, *25*, 193–197. [CrossRef]

8. Cárdenas, M.A.; García-López, D.; Gobernado-Mitre, I.; Merino, J.C.; Pastor, J.M.; de D. Martínez, J.; Barbeta, J.; Calveras, D. Mechanical and fire retardant properties of EVA/clay/ATH nanocomposites—Effect of particle size and surface treatment of ATH filler. *Polym. Degrad. Stab.* **2008**, *93*, 2032–2037.

9. Kuila, T.; Khanra, P.; Mishra, A.K.; Kim, N.H.; Lee, J.H. Functionalized-graphene/ethylene vinyl acetate co-polymer composites for improved mechanical and thermal properties. *Polym. Test.* **2012**, *31*, 282–289. [CrossRef]

10. Beyer, G. Flame retardant properties of organoclays and carbon nanotubes and their combination with alumina trihydrate. In *Flame Retardant Polymer Nanocomposites*; Morgan, A.B., Wilkie, C.A., Eds.; John Wiley & Sons, Inc.: Hoboken, NJ, USA, 2007; pp. 163–164.

11. Sauerwein, R. Mineral filler flame retardants. In *Non-Halogenated Flame Retardant Handbook*; Morgan, A.B., Wilkie, C.A., Eds.; Scrivener Publishing LLC: Salem, MA, USA, 2014; pp. 117–119.

12. Xu, B.; Ma, W.; Bi, X.; Shao, L.; Qian, L. Synergistic Effects of Nano-zinc Oxide on Improving the Flame Retardancy of EVA Composites with an Efficient Triazine-Based Charring Agent. *J. Polym. Environ.* **2019**, *27*, 1127–1140. [CrossRef]

13. Feng, C.; Liang, M.; Jiang, J.; Liu, H.; Huang, J. Synergistic effect of ammonium polyphosphate and triazine-based charring agent on the flame retardancy and combustion behavior of ethylene-vinyl acetate copolymer. *J Anal. Appl. Pyrolysis* **2016**, *119*, 259–269. [CrossRef]

14. Enescu, D.; Frache, A.; Lavaselli, M.; Monticelli, O.; Marino, F. Novel phosphorous-nitrogen intumescent flame retardant system. Its effects on flame retardancy and thermal properties of polypropylene. *Polym. Degrad. Stab.* **2013**, *98*, 297–305. [CrossRef]

15. Chen, H.; Wang, J.; Ni, A.; Ding, A.; Han, X.; Sun, Z. The Effects of a Macromolecular Charring Agent with Gas Phase and Condense Phase Synergistic Flame Retardant Capability on the Properties of PP/IFR Composites. *Materials* **2018**, *11*, 111. [CrossRef] [PubMed]

16. Parbhakar, A.; Cuadros, J.; Sephton, M.A.; Dubbin, W.; Coles, B.J.; Weiss, D. Adsorption of L-lysine on montmorillonite. *Coll. Surf. A Physicochem. Eng. Asp.* **2007**, *307*, 142–149. [CrossRef]

17. Cuadros, J.; Aldega, L.; Vetterlein, J.; Drickamer, K.; Dubbin, W. Reactions of lysine with montmorillonite at 80 °C: Implications for optical activity, H^+ transfer and lysine-montmorillonite binding. *J. Colloid Interface Sci.* **2009**, *333*, 78–84. [CrossRef] [PubMed]

18. Kitadai, N.; Yokoyama, T.; Nakashima, S. In situ ATR-IR investigation of L-lysine adsorption on montmorillonite. *J. Colloid Interface Sci.* **2009**, *338*, 395–401. [CrossRef]

19. Moly, K.A.; Bhagawan, S.S.; Groeninckx, G.; Thomas, S. Correlation Between the Morphology and Dynamic Mechanical Properties of Ethylene Vinyl Acetate/Linear Low-Density Polyethylene Blends: Effects of the Blend Ratio and Compatibilization. *J. Appl. Polym. Sci.* **2006**, *100*, 4526–4538. [CrossRef]

20. Utracki, L.A. Compatibilization of Polymer Blends. *Can. J. Chem. Eng.* **2002**, *80*, 1008–1016. [CrossRef]

21. Cogen, J.M.; Lin, T.S.; Whaley, P.D. Material Design for Fire Safety in Wire and Cable Applications. In *Fire Retardacy of Polymeric Materials*, 2nd ed.; Wilkie, C.A., Morgan, A.B., Eds.; CRC Press: Boca Raton, FL, USA, 2010; pp. 789–792.

22. Dennis, H.R.; Hunter, D.L.; Chang, D.; Kim, S.; White, J.L.; Cho, J.W.; Paul, D.R. Effect of melt processing conditions on the extent of exfoliation in organoclay-based nanocompostes. *Polymer* **2001**, *42*, 9513–9522. [CrossRef]

23. *Standard Test Method for Measuring the Comparative Burning Characteristics of Solid Plastics in a Vertical Position*; ASTM D3801-19; ASTM International: West Conshohocken, PA, USA, 2019.

24. *Standard Test Method for Tensile Properties of Plastics*; ASTM D638; ASTM International: West Conshohocken, PA, USA, 2014.

25. Cser, F.; Jollands, M.; White, P.; Bhattacharya, S. Miscibility studies on cross-linked EVA/LLDPE blends by TMDSC. *J. Therm. Anal. Calorim.* **2002**, *70*, 651–662. [CrossRef]

26. Khonakdar, H.A.; Jafari, S.H.; Yavari, A.; Asadinezhad, A.; Wagenknecht, U. Rheology, Morphology and Estimation of Interfacial Tension of LDPE/EVA and HDPE/EVA Blends. *Polym. Bull.* **2005**, *54*, 75–84. [CrossRef]

27. Ma, Y.; Qi, F.; Chen, M.; Chen, X.; Qin, J.; Tang, M. Study on Viscoelastic, Crystallization, and Mechanical Properties of HDPE/EVA/Mg(OH)$_2$ Composites. *Polym. Compos.* **2017**, *38*, 1221–1226. [CrossRef]

28. Salehiyan, R.; Ray, S.S.; Stadler, F.J.; Ojijo, V. Rheology–Microstructure Relationships in Melt-Processed Polylactide/Poly(vinylidene Fluoride) Blends. *Materials* **2018**, *11*, 2450. [CrossRef] [PubMed]

29. Bayram, H.; Önal, M.; Yılmaz, H.; Sarıkaya, Y. Thermal analysis of a white calcium bentonite. *J. Therm. Anal. Calorim.* **2010**, *101*, 73–879. [CrossRef]

30. Ursu, A.V.; Jinescu, G.; Gros, F.; Nistor, I.D.; Miron, N.D.; Lisa, G.; Silion, M.; Djelveh, G.; Azzouz, A. Thermal and chemical stability of Romanian bentonite. *J. Therm. Anal. Calorim.* **2011**, *106*, 965–971. [CrossRef]

31. Costa, L.; Camino, G.; Luda di Cortemiglia, M.P. Mechanism of thermal degradation of fire-retardant melamine salts. In *Fire and Polymers Hazards Identification and Prevention*; Nelson, G.L., Ed.; American Chemical Society: Washington, DC, USA, 1990; pp. 211–213.

32. Alongi, J.; Di Blasio, A.; Cuttica, F.; Carosio, F.; Malucelli, G. Flame Retardant Properties of Ethylene Vinyl Acetate Copolymers Melt-Compounded with Deoxyribonucleic Acid in the Presence of a-cellulose or beta-cyclodextrins. *Curr. Org. Chem.* **2014**, *18*, 1651–1660. [CrossRef]

33. Zanetti, M.; Camino, G.; Thomann, R.; Mülhaupt, R. Synthesis and thermal behaviour of layered silicate-EVA nanocomposites. *Polymer* **2001**, *42*, 4501–4507. [CrossRef]

34. Riva, A.; Zanetti, M.; Braglia, M.; Camino, G.; Falqui, L. Thermal degradation and rheological behaviour of EVA/montmorillonite nanocomposites. *Polym. Degrad. Stab.* **2002**, *77*, 299–304. [CrossRef]

35. Zanetti, M.; Bracco, P.; Costa, L. Thermal degradation behaviour of PE/clay nanocomposites. *Polym. Degrad. Stab.* **2004**, *85*, 657–665. [CrossRef]

36. Sanchez-Olivares, G.; Sanchez-Solis, A.; Calderas, F.; Medina-Torres, L.; Herrera-Valencia, E.E.; Castro-Aranda, J.I.; Manero, O.; Di Blasio, A.; Alongi, J. Flame retardant high density polyethylene optimized by on-line ultrasound extrusion. *Polym. Degrad. Stab.* **2013**, *98*, 2153–2160. [CrossRef]

37. Camino, G.; Maffezzoli, A.; Braglia, M.; De Lazzaro, M.; Zammarano, M. Effect of hydroxides and hydroxycarbonate structure on fire retardant effectiveness and mechanical properties in ethylene-vinyl acetate copolymer. *Polym. Degrad. Stab.* **2001**, *74*, 457–464. [CrossRef]

38. Hoffendahl, C.; Fontaine, G.; Duquesne, S.; Taschner, F.; Mezger, M.; Bourbigot, S. The combination of aluminum trihydroxide ATH) and melamine borate (MB) as fire retardant additives for elastomeric ethylene vinyl acetate (EVA). *Polym. Degrad. Stab.* **2015**, *115*, 77–88. [CrossRef]

39. Yang, W.; Hu, Y.; Tai, Q.; Lu, H.; Song, L.; Yuen, R.K.K. Fire and mechanical performance of nanoclay reinforced glass-fiber/PBT composites containing aluminum hypophosphite particles. *Compos. Part A Appl. Sci. Manuf.* **2011**, *42*, 794–800. [CrossRef]

40. Yang, W.; Kan, Y.; Song, L.; Hu, Y.; Lu, H.; Yuen, R.K.K. Effect of organo-modified montmorillonite on flame retardant poly(1,4-butylene terephthalate) composites. *Polym. Adv. Technol.* **2011**, *22*, 2564–2570. [CrossRef]

41. Sanchez-Olivares, G.; Sanchez-Solis, A.; Calderas, F.; Medina-Torres, L.; Herrera-Valencia, E.E.; Rivera-Gonzaga, A.; Manero, O. Extrusion with Ultrasound Applied on Intumescent Flame-Retardant Polypropylene. *Polym. Eng. Sci.* **2013**, *53*, 2018–2026. [CrossRef]

MDPI
St. Alban-Anlage 66
4052 Basel
Switzerland
Tel. +41 61 683 77 34
Fax +41 61 302 89 18
www.mdpi.com

Materials Editorial Office
E-mail: materials@mdpi.com
www.mdpi.com/journal/materials